THE
SCIENCE
OF
AIR

CONCEPTS
AND
APPLICATIONS

SECOND EDITION

THE
SCIENCE
OF
AIR

CONCEPTS
AND
APPLICATIONS
SECOND EDITION

FRANK R. SPELLMAN

CRC Press
Taylor & Francis Group
Boca Raton London New York

CRC Press is an imprint of the
Taylor & Francis Group, an **informa** business

CRC Press
Taylor & Francis Group
6000 Broken Sound Parkway NW, Suite 300
Boca Raton, FL 33487-2742

First issued in paperback 2019

ISBN-13: 978-1-4200-7532-8 (hbk)
ISBN-13: 978-0-367-38668-9 (pbk)

Library of Congress Cataloging-in-Publication Data

Spellman, Frank R.
    The science of air : concepts and applications / Frank R. Spellman. -- 2nd ed.
        p. cm.
    Includes bibliographical references and index.
    ISBN 978-1-4200-7532-8 (alk. paper)
    1. Atmospheric physics. 2. Atmospheric chemistry. I. Title.

QC861.3.S64 2009
551.51--dc22   2008032330

Visit the Taylor & Francis Web site at
http://www.taylorandfrancis.com

and the CRC Press Web site at
http://www.crcpress.com

Contents

PART I   Fundamentals

Chapter 3

Chapter 5

PART II Atmospheric Science

Chapter 6

Chapter 7

Chapter 12

PART III   Air Quality

Chapter 13

Chapter 14

Preface

Hailed on first publication as a masterly account for both the general reader and student, *The Science of Air: Concepts and Applications*, 2nd edition (a companion text to the highly successful *The Science of Water: Concepts and Applications*, 2nd edition) continues to deal with every aspect of air.

Central to this discussion about air and its critical importance on earth is man—man and his use, misuse, and reuse of air. This new edition takes the view that since air is the primary substance and sustenance of most life on earth, it is precious—too precious to abuse, pollute, and ignore. The common thread woven throughout the fabric of this text is air resource utilization and its protection.

In addition to asking the same questions, this standard synthesis has now been completely revised and expanded for the second edition—every section has been updated, with major expansions on upgrades in air math, air physics, and airflow parameters, with particular emphasis on indoor air quality. The text still deals with the essence of air, that is, what air is all about. Further, while this text points out that air is one of the simplest and most common chemical compounds on earth, it also shows air to be one of the most mysterious substances we know.

The text follows a pattern that is nontraditional; that is, the paradigm (model or prototype) used here is based on real-world experience—not on theoretical gobble-dygook. Clearly written and user-friendly, this timely update of *The Science of Air* builds on the remarkable success of the first edition. Still written as an information source, it should be pointed out that this text is not limited in its potential for other uses. For example, while this work can be utilized by the air pollution control practitioner/specialist to provide valuable insight into the substance he or she works hard to collect, treat, and resupply for its intended purpose, it can just as easily provide important information for the government and industrial policymaker who may be tasked with making decisions concerning air protection policies. Consequently, this book will serve a varied audience: students, lay personnel, regulators, technical experts, attorneys, business leaders, and concerned citizens.

The question becomes: Why a text on the science of air? The study of air is a science. It is a science that is closely related/interrelated to other scientific disciplines, such as biology, microbiology, chemistry, mathematics, hydrology, and others. Therefore, to solve the problems and understand the issues related to air, air pollution control practitioners need a broad base of scientific information from which to draw—*The Science of Air* fills this critical need.

Frank R. Spellman

Norfolk, Virginia

The Author

Frank R. Spellman is assistant professor of environmental health at Old Dominion University, Norfolk, Virginia, and author of 55 books. Spellman's books include *Concentrated Animal Feeding Operations* (CAFOs) and several others on topics in all areas of environmental science and occupational health. Many of Spellman's texts are listed on Amazon.com and Barnes and Noble. Several of his texts have been adopted for classroom use at major universities throughout the United States, Canada, Europe, and Russia; two are currently being translated in Spanish for South American markets.

Spellman has been cited in more than 400 publications, serves as a professional expert witness for three law groups, is an accident investigator for a Northern Virginia law firm, and consults on Homeland Security vulnerability assessments (VAs) for critical infrastructure, including water/wastewater facilities nationwide.

Spellman receives numerous requests to coauthor with well-recognized experts in several scientific fields. For example, he is a contributing author for the prestigious text *The Engineering Handbook*, 2nd edition (CRC Press).

Spellman lectures on sewage treatment, water treatment, homeland security, and health and safety topics throughout the country and teaches water/wastewater operator short courses at Virginia Tech (Blacksburg).

Spellman holds a BA in public administration, a BS in business management, an MBA, an MS in environmental engineering, and a PhD in environmental engineering.

Part I

Fundamentals

1 Introduction

Whether we characterize it as a caress, a gentle breeze, or a warm wind, or a blustery gale, tempest, typhoon, tornado, or hurricane, air is vital. Air can dry, cool, warm, freshen, ventilate, or irritate us.

Air is scientifically unique. The combination of common and rare gases we breathe has made life possible. Air, as with water, is the only chemical compound found naturally that affects most living organisms in a manner of ways. We cannot imagine life without breathing—we must constantly quench our "thirst" for air.

Air sustains growth. It creates the subtle and blatant movements that provide us with changing weather patterns. Air gives us the blessing of communication. From our first cry to our dying breath, our voices travel on a current of air.

Pure air is odorless, colorless, and tasteless. We rarely stop to think about it unless it brings something to us as a reminder. It covers earth completely; nothing can escape air's touch. Air is life—life and air are inseparable. We sometimes call air the breath of life—a fitting name. But polluted air can become the bane of life as we know it.

Whether it provides the fundamental source of power in a sailing vessel or pushes the blades of a windmill, a billowy cloud, a dust mote, or a feather; whether it lifts a bird soaring on thermals; whether it wafts to us the sweet fragrance of gardenia, lavender, lilac, or rose—or a seed to fertile ground; whether it sets water lapping against some distant shore, drives a gritty wind that sculpts mountains to sand, or hammers a horrendous fist that flattens whatever stands in its path—cities, forests, crops, and man—air is essential.

Our very existence depends on air, but we have created a paradox with our vital line to life. Why would we abuse something so vital, something we cannot live without? Why do we foul the very essence of our lives? Why do we insult our environment at a faster pace than we can understand and mitigate the consequence? Why? Because air is everywhere. We have always had enough.

Let us hope that we always will. Let us hope that we are not destroying the very air we breathe. Let us hope that technology will aid us in our efforts to retain the quality of air we need to survive.

We need air as it should be: unpolluted and in the perfect mixture of elements we were evolved to inhale. Air does not care about the greenhouse effect, ozone depletion, or global warming, but we should. The bottom line is that we must not forget that we exist by environmental consent, subject to change without notice.

F. Spellman

SETTING THE STAGE

What is air? Most of us would have little difficulty in quickly answering this question by stating that air is the oxygen that we breathe—the substance that we need to sustain life.

3

Taking our definition of air to the next level and beyond, we might state that air surrounds us—air is virtually everywhere; air is that substance we feel against our faces and skin when the wind blows; air is that gas that we use to fill our automobile tires; air is necessary for combustion to take place; hot air lifts our balloons; air under pressure (pneumatic air) powers our tools and machines; air can be either warm or cold or just about right; air is air—what more needs to be said? A lot more.

What is air? The environmental scientist/practitioner would answer this question differently than most of us. He or she would state that air is a gas—actually, a combination of gases. The scientist might also say that a gas is a state of matter distinguished from the solid and liquid states by very low density and viscosity, relatively great expansion and contraction with changes in pressure and temperature, the ability to diffuse readily, and the spontaneous tendency to become distributed throughout any container.

How about the engineer? Engineers always seem to have a definition for just about anything and everything (most of which cannot be understood by many of us—and maybe that is their intent). An engineer might refer to air as a fluid (because it is; like water, air is fluid—you can pour it). Engineers are primarily interested in air as a fluid because they deal with fluid mechanics, the study of the behavior of fluids (including air) at rest or in motion. Fluids may be either gases or liquids. You are probably familiar with the physical difference between gases and liquids, as exhibited by air and water, but for the study of fluid mechanics (and the purposes of this text), it is convenient to classify fluids based on their compressibility:

- Gases are very readily compressible (you have heard of compressed air).
- Liquids are only slightly compressible (it is unlikely you have heard much about compressed water).

What is air? Air is a mixture of gases that constitutes the earth's atmosphere. What is the earth's atmosphere? The atmosphere is that thin shell, veil, an envelope of gases that surrounds earth like the skin of an apple—very thin, but very vital. The approximate composition of dry air is, by volume at sea level, nitrogen 78%, oxygen 21% (necessary for life as we know it), argon 0.93%, and carbon dioxide 0.03%, together with very small amounts of numerous other constituents (see Table 1.1). Because of constant mixing by the winds and other weather factors, the percentages of each gas in the atmosphere are normally constant to 70,000 feet. However, it is important to point out that the water vapor content is highly variable and depends on atmospheric conditions. Air is said to be pure when none of the minor constituents is present in sufficient concentration to be injurious to the health of human beings or animals, to damage vegetation, or to cause loss of amenity (e.g., through the presence of dirt, dust, or odors, or by diminution of sunshine).

Where does air come from? Genesis 1:2 states that God separated the water environment into the atmosphere and surface waters on the second day of creation. Many scientists state that 4.6 billion years ago a cloud of dust and gases forged the earth and also created a dense molten core enveloped in cosmic gases. This was

TABLE 1.1
Composition of Air/Earth's Atmosphere

Gas	Chemical Symbol	Volume (%)
Nitrogen	N_2	78.08
Oxygen	O_2	20.94
Carbon dioxide	CO_2	0.03
Argon	Ar	0.093
Neon	Ne	0.0018
Helium	He	0.0005
Krypton	Kr	Trace
Xenon	Xe	Trace
Ozone	O_3	0.00006
Hydrogen	H_2	0.00005

the *proto-atmosphere* or *proto-air*, composed mainly of carbon dioxide, hydrogen, ammonia, and carbon monoxide, but it was not long before it was stripped away by a tremendous outburst of charged particles from the sun. As the outer crust of earth began to solidify, a new atmosphere began to form from the gases outpouring from gigantic hot springs and volcanoes. This created an atmosphere of air composed of carbon dioxide, nitrogen oxides, hydrogen, sulfur dioxide, and water vapor. As the earth cooled, water vapor condensed into highly acidic rainfall, which collected to form oceans and lakes.

For much of earth's early existence (the first half), only trace amounts of free oxygen were present. But then green plants evolved in the oceans and they began to add oxygen to the atmosphere as a waste gas, and later oxygen increased to about 1% of the atmosphere, and with time to its present 21%.

How do we know for sure about the evolution of air on earth? Are we just guessing, using "voodoo" science? There is no guessing or voodoo involved with the historical geological record. Consider, for example, geological formations that are dated to 2 billion years ago. In these early sediments there is a clear and extensive band of red sediment (red bed sediments)—sands colored with oxidized (ferric) iron. Previously, ferrous formations had been laid down showing no oxidation. But there is more evidence. We can look at the timeframe of 4.5 billion years ago, when carbon dioxide in the atmosphere was beginning to be lost in sediments. The vast amount of carbon deposited in limestone, oil, and coal indicates that carbon dioxide concentrations must once have been many times greater than they are today, only 0.03%. The first carbonated deposits appeared about 1.7 billion years ago, and the first sulfate deposits about 1 billion years ago. The decreasing carbon dioxide was balanced by an increase in the nitrogen content of the air. The forms of *respiration* practiced advanced from fermentation 4 billion years ago to anaerobic *photosynthesis* 3 billion years ago to aerobic photosynthesis 1.5 billion years ago. The aerobic respiration that is so familiar today only began to appear about 500 million years ago.

Did You Know?

Aerobic respiration is the release of energy from glucose or another organic sub-strate in the presence of oxygen. Strictly speaking, *aerobic* means "in air," but it is the oxygen in the air that is necessary for aerobic respiration.

Fast-forward to the present. The atmosphere itself continues to evolve, but human activities—with their highly polluting effects—have now overtaken nature in deter-mining the changes. And that is the overriding theme of this text: human beings and their effect on earth's air.

Have you ever wondered where the air goes when we expel it from our lungs, or if, when we do so, it is still air? When we use air to feed our fires, power our machines, weld or braze our metals, vacuum our floors, spray our propellants—paints, insecticides, lubricants, etc.—do we change the nature of air? Have you ever really wondered about these things? Probably not. Are they important? Maybe. Maybe not. Are these questions and their answers important in this text? Yes. Why? This text is about the science of air, and thus all things that affect air are important here.

At this point you are probably asking yourself: What does all this have to do with anything? Good question. What it has to do with air is quite simple. We do not know as much as we need to know about air. As a case in point, consider this: Have you ever gone to the library and tried to find a text that deals exclusively and extensively with the science of air? Such texts are few, far-flung, imaginary, nonexistent—a huge information gap out there.

To start with, let us talk about air—about breathing air in particular (this text will discuss all aspects of air—breathing air and working air), the air we need to survive, to sustain life, the air that probably concerns us the most.

When the average person takes in a deep breath of air, he or she probably gives little thought to what he or she is doing, that is, breathing life-sustaining air. Let's face it, breathing a breath of air is something that normally takes little effort and even less thought. The situation could be different, however. For example, consider a young woman, a firefighter—an emergency services provider. On occasion she has to fight fires where she must wear a self-contained breathing apparatus (SCBA) to avoid breathing smoke and decreased oxygen levels created by the fire. The standard SCBA with a single bottle contains approximately 45 minutes of air (class D breath-ing air, which is not oxygen but regular air with 21% oxygen and associated gases, nitrogen, etc.).

On this particular day, our firefighter responds to a fire where she and another firefighter are required to enter a burning building to look for and rescue any trapped victims. To enter such a building, the firefighters don their SCBAs, activate their air supply, and enter the burning structure. Normally, 45 minutes of air is plenty to make a quick survey of a house's interior, especially when it is on fire. However, sometimes (this was one of them) the best laid plans go awry.

After having swept the first floor of the two-story house without discovering any victims, the two firefighters climb the stairs to the second floor to look. But the fire, which started in the kitchen, is spreading fast, and the smoke and toxic vapors are spreading even faster. The firefighters know this—they are well-trained professionals. They know that any person within this house without respiratory protection will not survive for long. The fire will not kill them, but the smoke and toxic vapors surely will.

At the landing upstairs, the firefighters are on their knees, crawling, looking for victims. The smoke and toxic vapors are intense and intensifying by the second. But the firefighters are not worried (not yet); they have all the air they need strapped to their backs.

By the time they reach the hallway, the visibility is zero, the heat intense, and the toxic vapors and smoke so thick you cannot see them but can feel them. All is well until flames find their way up the stairs and quickly spread down the carpeted hallway to the backs of the firefighters. They have 15 minutes of air left.

But then the flames are becoming intimate; they reach out and touch the firefighters. The situation has instantly changed from one of rescuing victims to one of fleeing for their own lives. They have 12 minutes of air left.

They about-face on their hands and knees—and face the fire. Their only hope of escape is through the flames, but they are not too worried—they are well equipped with fire protective clothing and have 9 minutes of air left.

Normally, 9 minutes of air is a lot in most escape situations. But this was not a normal situation. As the firefighters slid down the stairs to the first floor, their air supply registered 2 minutes. They had used more air in the last 30 seconds than they had in the previous 10 minutes. This excessive use of air should come as no surprise when you consider that with toxic vapors and smoke and flames all over the place, and the flames licking at your body, you too might have a tendency to get excited, to breath in and out copious amounts of air. Our two firefighters were no different than you or me—they were scared and breathed hard until they breathed their last. They fell unconscious right in front of the doorway—just one more breath of air with 21% oxygen and they would have escaped. But they did not. The irony is that the fire—well beyond its flashover state—had all the air, with its accompanying oxygen supply, it needed to continue its deadly destruction.

The preceding example sheds light on a completely different view of air. Actually, a very basic view that holds: We cannot live without it.

Did You Know?

Because of an actual incident similar to the one described above that caused two firefighters to lose their lives fighting a structure fire, OSHA's 29 CFR 1910.134 Respiratory Protection standard reemphasized firefighters' requirement to abide by the standard. Specifically, OSHA stipulates the two in/two out rule. This rule ensures that the two in can monitor each other and assist with equipment failure or entrapment or other hazards, and the two out can monitor those in the building, initiate rescue, or call for backup. One of the two out can be assigned another role, such as incident commander. Moreover, under OSHA's 134 standard, firefighters (and anyone else using respirators) must check the condition of the respirator before donning it for use.

TABLE 1.2
Mortality Occurring during Air Pollution
Events

Location	Year	Deaths Reported as a Result of Pollution Event
Belgium	1930	63
Pennsylvania	1948	17
London	1948	700–800
London	1952	4,000
London	1956	1,000
London	1957	700–800
London	1959	200–250
London	1962	700
London	1963	700
New York	1963	200–400

Source: Subcommittee on Air and Water Pollution (1968).

If we cannot live without air—if air is so necessary for sustaining life—then two questions arise: (1) Why do we ignore air? (2) Why do we abuse it (pollute it)? We ignore air (like we do water) because it is so common, so accessible (normally), so unexceptional (unless you are in a burning building and your life depends on it). Why do we pollute air? There are several reasons; many will be discussed later in this text.

You might be asking yourself: Is air pollution really that big of a deal? Isn't pollution relative? That is, isn't pollution dependent on your point of view—a judgment call? Well, if you could ask the victims of the incidents listed in Table 1.2, the answer, simply stated, would be yes, it is a big deal.

Air is one of our essential resources, sustaining life as it stimulates and pleases the senses—though invisible to the human eye, it makes possible such sights as beautiful and dazzling rainbows, heart-pinching sunsets and sunrises, the Northern Lights, and on occasion, a clear view of that high alpine meadow sprinkled throughout with the colors of spring. But air is more than this. Air is capable of many other wondrous things. For example, have you ever felt the light touch of a cool soothing breeze against your skin? Yet air is capable of much more: it carries thousands of scents—both pungent and subtle: salty ocean breezes, approaching rain, fragrances from blooming flowers, and others.

It is the "others" that concern us here: the sulfurous gases from industrial processes (that typical rotten egg odor); the stink of garbage, refuse, trash (all part of humankind's throwaways); the toxic poison remnants from pesticides, herbicides, and all the other "cides." We are surrounded by it but seldom think about it until it displeases us (remember, humans can put up with just about anything until it displeases them). It is pollution, those discarded, sickening leftovers of the heavy hand of humans, that causes the problem. As stated previously, we will cover this life-threatening travesty of polluting our environment in greater detail later in the text.

Keep in mind that as this text proceeds it leads us down a path sometimes heavy with soot, chemicals, smoke, malodorous scents, and particulate matter, but at all times we will progress with the sense of importance that such a simple substance such as air, containing only a few gaseous elements and other products, has on our lives.

In the opening it was stated that a gap in knowledge exists when it comes to dealing with the science of air. This text is designed to bridge the gap, to fill in this obvious and unsatisfactory gap in information about air, because on this planet, air is life.

SCOPE OF TEXT

This text aligns itself with the same preamble stated in its two sister texts, *The Science of Water* and *The Science of Environmental Pollution*: science is any systematic field of study or body of knowledge that aims, through experiment, observation, and deduction, to produce reliable explanation of phenomena, with reference to the material and physical world.

It is the intent of this text to abide by the precepts presented in this preamble. Moreover, this text presents the science of air in a logical, step-by-step, plain English, user-friendly format.

Having said all this, it should be pointed out that this text is specifically written to serve as a reference for an undergraduate course and also as a guide to prepare air pollution control technologists for practice in the real world. It also serves as a basic primer for those who are interested in gaining knowledge about air-related topics, for whatever reason. The text follows a pattern that is nontraditional. The paradigm (model or prototype) used here is based on real-world experience and application—not on theoretical gobbledygook. It is the knowledge gained from work performed in the world of environmental science that this text is all about.

DEFINITION OF KEY TERMS

Every branch of science, including air science, has its own language for communication. The terminology used herein is as different as that from aeronautical engineering is from agronomy's. In order to work at even the edge of air science and the science disciplines closely related to air science, it is necessary for the reader to acquire a familiarity with the vocabulary used in this text.

While it is helpful and important for technical publications to include definitions or a glossary of key terms at the end of the work, for the reader's use, it is more important, in my view, to include many of these key definitions early in the text to facilitate a more orderly, logical, step-by-step learning activity. Thus, below some of the key terms are listed and defined. Other terms not defined here will be defined when they are used in the text.

Absolute pressure—The total pressure in a system, including both the pressure of a substance and the pressure of the atmosphere (about 14.7 psi, at sea level).

Acid—Any substance that releases hydrogen ions (H^+) when it is mixed into water.

Acid precipitation—Rain, snow, or fog that contains higher than normal levels of sulfuric or nitric acid, which may damage forests, aquatic ecosystems, and cultural landmarks.

Acid surge—A period of short, intense acid deposition in lakes and streams as a result of the release (by rainfall or spring snowmelt) of acids stored in soil or snow.

Acidic solution—A solution that contains significant numbers of H^+ ions.

Airborne toxins—Hazardous chemical pollutants that have been released into the atmosphere and are carried by air currents.

Albedo—Reflectivity, or the fraction of incident light that is reflected by a surface.

Arithmetic mean—A measurement of average value, calculated by summing all terms and dividing by the number of terms.

Arithmetic scale—A scale is a series of intervals (marks or lines), usually made along the side or bottom of a graph, that represents the range of values of the data. When the marks or lines are equally spaced, it is called an arithmetic scale.

Atmosphere—A 500-kilometer-thick layer of colorless, odorless gases known as air that surrounds the earth and is composed of nitrogen, oxygen, argon, carbon dioxide, and other gases in trace amounts.

Atom—The smallest particle of an element that still retains the characteristics of that element.

Atomic number—The number of protons in the nucleus of an atom.

Atomic weight—The sum number of protons and the number of neutrons in the nucleus of an atom.

Base—Any substance that releases hydroxyl ions (OH^-) when it dissociates in water.

Chemical bond—The force that holds atoms together within molecules. A chemical bond is formed when a chemical reaction takes place. Two types of chemical bonds are ionic bonds and covalent bonds.

Chemical reaction—A process that occurs when atoms of certain elements are brought together and combine to form molecules, or when molecules are broken down into individual atoms.

Climate—The long-term weather pattern of a particular region.

Covalent bond—A type of chemical bond in which electrons are shared.

Density—The weight of a substance per unit of its volume, e.g., pounds per cubic foot.

Dew point—The temperature at which a sample of air becomes saturated, i.e., has a relative humidity of 100%.

Element—Any of more than 100 fundamental substances that consist of atoms of only one kind and that constitute all matter.

Emission standards—The maximum amount of a specific pollutant permitted to be legally discharged from a particular source in a given environment.

Emissivity—The relative power of a surface to reradiate solar radiation back into space in the form of heat, or long-wave infrared radiation.

Energy—The ability to do work, to move matter from place to place, or to change matter from one form to another.

First law of thermodynamics—Natural law that dictates that during physical or chemical change energy is neither created nor destroyed, but it may be changed in form and moved from place to place.

Global warming—The increase in global temperature predicted to arise from increased levels of carbon dioxide, methane, and other greenhouse gases in the atmosphere.

Greenhouse effect—The prevention of the reradiation of heat waves to space by carbon dioxide, methane, and other gases in the atmosphere. The greenhouse effect makes possible the conditions that enable life to exist on Earth.

Insolation—The solar radiation received by the earth and its atmosphere—incoming solar radiation.

Ion—An atom or radical in solution carrying an integral electrical charge, either positive (cation) or negative (anion).

Lapse rate—The rate of temperature change with altitude. In the troposphere the normal lapse rate is –3.5°F per 1,000 ft.

Matter—Anything that exists in time, occupies space, and has mass.

Mesosphere—A region of the atmosphere based on temperature between approximately 35 to 60 miles in altitude.

Meteorology—The study of atmospheric phenomena.

Mixture—Two or more elements, compounds, or both, mixed together with no chemical reaction occurring.

Ozone—The compound O_3. It is found naturally in the atmosphere in the ozonosphere and is also a constituent of photochemical smog.

pH—A means of expressing hydrogen ion concentration in terms of the powers of 10; measurement of how acidic or basic a substance is. The pH scale runs from 0 (most acidic) to 14 (most basic). The center of the range (7) indicates the substance is neutral.

Photochemical smog—An atmospheric haze that occurs above industrial sites and urban areas resulting from reactions, which take place in the presence of sunlight, between pollutants produced in high temperature and the pressurized combustion process (such as the combustion of fuel in a motor vehicle). The primary component of smog is ozone.

Photosynthesis—The process of using the sun's light energy by chlorophyll-containing plants to convert carbon dioxide (CO_2) and water (H_2O) into complex chemical bonds forming simple carbohydrates such as glucose and fructose.

Pollutant—A contaminant at a concentration high enough to endanger the environment.

Pressure—The force pushing on a unit area. Normally, in air applications, measured in atmospheres, pascal (Pa) or pounds per square inch (psi).

Primary pollutants—Pollutants that are emitted directly into the atmosphere where they exert an adverse influence on human health of the environment. The six primary pollutants are carbon dioxide, carbon monoxide, sulfur oxides, nitrogen oxides, hydrocarbons, and particulates. All but carbon dioxide are regulated in the United States.

Raleigh scattering—The preferential scattering of light by air molecules and particles that accounts for the blueness of the sky. The scattering is proportional to $1/\lambda^4$.

Radon—A naturally occurring radioactive gas, arising from the decay of uranium 238, which may be harmful to human health in high concentrations.

Rain shadow effect—The phenomenon that occurs as a result of the movement of air masses over a mountain range. As an air mass rises to clear a mountain, the air cools and precipitation forms. Often, both the precipitation and the pollutant load carried by the air mass will be dropped on the windward side of the mountain. The air mass is then devoid of most of its moisture; consequently, the lee side of the mountain receives little or no precipitation and is said to lie in the rain shadow of the mountain range.

Relative humidity—The concentration of water vapor in the air. It is expressed as the percentage that its moisture content represents the maximum amount that the air could contain at the same temperature and pressure. The higher the temperature, the more water vapor the air can hold.

Secondary pollutants—Pollutants formed from the interaction of primary pollutants with other primary pollutants or with atmospheric compounds such as water vapor.

Second law of thermodynamics—Natural law that dictates that with each change in form some energy is degraded to a less useful form and given off to the surroundings, usually as low-quality heat.

Solute—The substance dissolved in a solution.

Solution—A liquid containing a dissolved substance.

Specific gravity—The ratio of the density of a substance to a standard density. For gases, the density is compared with the density of air (= 1).

Stratosphere—An atmospheric layer extending from 6 to 7 miles to 30 miles above the earth's surface.

Stratospheric ozone depletion—The thinning of the ozone layer in the stratosphere; occurs when certain chemicals (such as chlorofluorocarbons) capable of destroying ozone accumulate in the upper atmosphere.

Thermosphere—An atmospheric layer that extends from 56 miles to outer space.

Troposphere—The atmospheric layer that extends from the earth's surface to 6 to 7 miles above the surface.

Weather—The day-to-day pattern of precipitation, temperature, wind, barometric pressure, and humidity.

Wind—Horizontal air motion.

REFERENCES AND RECOMMENDED READING

EPA. 2007. Introduction to air pollution. http://www.epa.gov/air/oaqps/eog/course42/ap.html (accessed December 25, 2007).

NASA. 2007. *The water planet.* http://rst.gsfc.nasa.gov/Sect14/Sect14_1b.html (Accessed 12/25/07).

Spellman, F. R. 1998. *Environmental science and technology: Concepts and applications.* Lancaster, PA: Technomic Publishing Company.

Spellman, F. R., and Whiting, N. 2006. *Environmental science and technology: Concepts and applications.* 2nd ed. Rockville, MD: Government Institutes.

Subcommittee on Air and Water Pollution, Committee on Public Works, U.S. Senate. 1968. *Air quality criteria.* Staff report 94-411, Author.

2 All about Air

When informed that air is a predominantly mechanical mixture of a variety of individual gases (most of which will not directly sustain life) forming the earth's enveloping atmosphere, many students and the uninitiated really do not give this information much thought. Why should they? These words usually fall on (if polluted) air, and that's that. So, as with water and soil, air is just one of those substances that we all know we need, but that is as far as our thoughts generally go on the topic.

Our early ancestors may have viewed air in a different light; the historical record seems to support this. For even though they were ignorant about the nuts and bolts of science and nature, they understood the import of the sun, water, soil, and air. Evidence indicates that they paid homage to gods for each of these life-sustaining substances, because they, unlike most of us, understood that these substances were a blessing and were to be held in the highest regard. When you get right down to it, this makes sense because they understood that life was not possible if any of these substances was lacking, nonexistent, or damaged beyond repair. Over much more time—millennia, in fact—we have come to recognize, to a degree, what our ancestors knew with surety from the very start: the sun, water, soil, and air are fundamental to everything we know—to everything we will ever know. One fact is absolutely certain: if we cannot find air to breathe, the other three vital substances are no longer important to us.

INTRODUCTION

As implied in the chapter opening, air is something that we do not normally think about unless we have to: we are short on air, or the air we breathe makes us cough, sick, or worse. When forced to think or talk about air, we generally do not think or talk about air; instead, we think or talk about the oxygen in the air. Let's face it, it is the gaseous oxygen that sustains our lives via respiration where the oxygen we take in burns the carbon in foodstuffs (like combustion), which then is exhaled; thus, the other gaseous and particulate substances that make up air are just along for the ride, right? Not exactly. It is true that air and oxygen have become synonymous—one in the same thing—to most people. Those of us who study or practice science know that air and oxygen are not synonymous; they are different, actually unique entities that just happen to blend or mix. From an earthly life-form point of view, this is a good thing. No doubt about it. Don't you agree?

In this chapter we talk about air—all about air. Not just the gaseous oxygen in air, but the natural gaseous and suspended minute liquids or solid particulates (aerosols) that make up air. Moreover, even though this text is primarily directed toward a discussion about atmospheric air and the environmental implications affecting it, this chapter also addresses other important functions air provides (besides the need to breathe): combustion and pneumatic power.

REVOLUTIONARY SCIENCE

Today almost every elementary school child can explain in basic terms the composition of air (and water). Most young children understand that the air we breathe contains oxygen, nitrogen, and other gases. There was a time (just a few hundred years ago), however, when the actual composition of air and water was nothing more than speculation.

The French aristocrat Antoine Lavoisier (1743–1794) is universally regarded as the founder (father) of modern chemistry. This lofty title was bestowed on Lavoisier for his great experiments and discoveries related to the major components that make up air (oxygen and nitrogen) and to a lesser degree for identifying the components of water (hydrogen and oxygen).

Most of Lavoisier's experiments and discoveries took place in the years just preceding the French Revolution. And even though he ranks high up there with the other great scientists of his time, Lavoisier was guillotined, on trumped-up charges, during the French Revolution in 1794. Joseph Lagrange (1736–1813), the great French mathematician, said: "It required only a moment to sever his head, and probably one hundred years will not suffice to produce another like it."

Lagrange's eulogy concerning Lavoisier and his scientific accomplishments is quite fitting. Why? What was so difficult about discovering the basic components of air, water, and the oxygen theory of combustion? That which seems so simple and elementary to us today was not so clear 200 years ago. Indeed, these discoveries made at that time were quite difficult. We must remember that in Lavoisier's time so-called chemists had no clear idea of what a chemical element was, nor any understanding of the nature of gases.

Lavoisier's discoveries were built on the works of others who preceded him or who were working on similar experiments during his lifetime. Lavoisier's work also provided a foundation for scientific discoveries that followed. For example, Lavoisier experimented with the findings of the German chemist Georg Stahl (1660–1743) and disproved them. Stahl proposed a theory that a combustible material burned because it contained a substance called *phlogiston* (charcoal is a prime example). Stahl knew that metallurgists obtained some metals from their ores by heating them with charcoal, which seemed to support the phlogiston theory of combustion. However, Lavoisier, in 1774, with the help of Joseph Priestley (1733–1804), proved that the phlogiston theory was wrong. Priestly had heated a clax (in this particular case, the burned residue of oxide of mercury) in a closed apparatus and collected the gas liberated in the process. Priestley had discovered that this gas supported combustion better than air.

Lavoisier repeated Priestley's experiments and convinced himself of the presence in air of a gas that combined with substances when they burn, and that it was the same gas given off when the oxide of mercury was heated. Thus, he proved that when a substance burned it combined with the oxygen in the air. He named this gas *oxygine*, or "acid former" (from the Greek), because he believed all acids contained oxygen.

In the meantime, Lavoisier had identified the other main component of air, nitrogen, which he named *azote*, from the Greek for "no life." He also demonstrated that when hydrogen, which chemists of the day called inflammable air, was burned with oxygen, water was formed.

Lavoisier restructured chemistry and gave it its modern form. His work provided a firm foundation for the atomic theory proposed by British chemist and physicist

John Dalton, and his elements were later classified in the periodic table. Lavoisier's work set the stage for the discovery of the other gaseous constituents in air, made later by other scientists.

Interesting Point

Justus von Liebig, in "Letters on Chemistry," No. 3, has this to say about Lavoisier: "He discovered no new body, no new property, no natural phenomenon previously unknown; but all the facts established by him were the necessary consequences of the labors of those who preceded him. His merit, his immortal glory, consists in this—that he infused into the body of the science a new spirit; but all the members of that body were already in existence, and rightly joined together."

With nitrogen and oxygen already identified as the primary constituents in air, and later carbon dioxide, water vapor, helium, ozone, and particulate matter, it was some time before the other gaseous constituents were identified. Argon was discovered in 1894 by British chemists John Rayleigh and William Ramsay after all oxygen and nitrogen had been removed chemically from a sample of air. Ramsay along with Englishman Morris Travers discovered neon. They also discovered krypton and xenon in 1889.

THE COMPONENTS OF AIR: CHARACTERISTICS AND PROPERTIES

It was pointed out that air is a combination of component parts: gases (see Table 1.1) and other matter (suspended minute liquid or particulate matter). In this section we discuss each of these components.

Note: Much of the information pertaining to atmospheric gases that follows was adapted from the Compressed Gas Association's *Handbook of Compressed Gases* (1990) and *Environmental Science and Technology: Concepts and Applications* (2006).

ATMOSPHERIC NITROGEN

Nitrogen (N_2) makes up the major portion of the atmosphere (78.03% by volume, 75.5% by weight). It is a colorless, odorless, tasteless, nontoxic, and almost totally inert gas. Nitrogen is nonflammable, will not support combustion, and is not life supporting. Not life supporting? No, gaseous nitrogen is not. The obvious question becomes: If gaseous nitrogen does not support life, what is it doing in our atmosphere—what good is it? Logical question. However, the question is incorrect; it implies something that is not true: nitrogen is indeed good and more. Without nitrogen, we could not survive.

Nitrogen is part of earth's atmosphere primarily because, over time, it has simply accumulated in the atmosphere and remained in place and in balance. This nitrogen accumulation process has occurred because, chemically, nitrogen is not very reactive. When released by any process, it tends not to recombine with other elements and accumulates in the atmosphere. And this is a good thing, because we need nitrogen. No, we do not need it for breathing, but we need it for other life-sustaining processes.

Let us take a look at a couple of reasons why gaseous nitrogen is so important to us. Although nitrogen in its gaseous form is of little use to us, after oxygen, carbon, and hydrogen, it is the most common element in living tissues. As a chief constituent of chlorophyll, amino acids, and nucleic acids—the building blocks of proteins (which are used as structural components in cells)—nitrogen is essential to life. Nitrogen is dissolved in and is carried by the blood. Nitrogen does not appear to enter into any chemical combination as it is carried throughout the body. Each time we breathe, the same amount of nitrogen is exhaled as is inhaled. Animals cannot use nitrogen directly but only when it is obtained by eating plant or animal tissues; plants obtain the nitrogen they need when it is in the form of inorganic compounds, principally nitrate and ammonium.

Gaseous nitrogen is converted to a form usable by plants (nitrate ions) chiefly through the process of nitrogen fixation via the nitrogen cycle, shown in simplified form in Figure 2.1.

Via the *nitrogen cycle*, aerial nitrogen is converted into nitrates mainly by microorganisms, bacteria, and blue-green algae. Lightning also converts some aerial nitrogen gas into forms that return to the earth as nitrate ions in rainfall and other types of precipitation. From Figure 2.1 it can be seen that ammonia plays a major role in the nitrogen cycle. Excretion by animals and anaerobic decomposition of dead organic matter by bacteria produce ammonia. Ammonia, in turn, is converted by nitrification bacteria into nitrites and then into nitrates. This process is known as nitrification. Nitrification bacteria are aerobic. Bacteria that convert ammonia into nitrites are known as nitrite bacteria (*Nitrosococcus* and *Nitrosomonas*). Although nitrite is toxic to many plants, it usually does not accumulate in the soil. Instead,

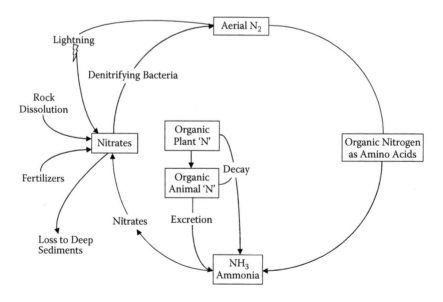

FIGURE 2.1 Nitrogen cycle.

other bacteria (such as *Nitrobacter*) oxidize the nitrite to form nitrate (NO_3^-), the most common biologically usable form of nitrogen.

Nitrogen reenters the atmosphere through the action of denitrifying bacteria, which are found in nutrient-rich habitats such as marshes and swamps. These bacteria break down nitrates into nitrogen gas and nitrous oxide (N_2O), which then reenter the atmosphere. Nitrogen also reenters the atmosphere from exposed nitrate deposits, emissions from electric power plants and automobiles, and from volcanoes.

Physical Properties

The physical properties of nitrogen are noted in Table 2.1.

Uses

In addition to being the preeminent (in regards to volume) component of earth's atmosphere and providing an essential ingredient in sustaining life, nitrogen gas has many commercial and technical applications. As a gas, it is used in heat treating primary metals; the production of semiconductor electronic components, as a blanketing atmosphere; blanketing of oxygen-sensitive liquids and volatile liquid chemicals; inhibition of aerobic bacteria growth; and the propulsion of liquids through canisters, cylinders, and pipelines.

Nitrogen Oxides

There are six oxides of nitrogen: nitrous oxide (N_2O), nitric oxide (NO), dinitrogen trioxide (N_2O_3), nitrogen dioxide (NO_2), dinitrogen tetroxide (N_2O_4), and dinitrogen pentoxide (N_2O_5).

Nitric oxide, nitrogen dioxide, and nitrogen tetroxide are fire gases. One or more of them is generated when certain nitrogenous organic compounds (polyurethane) burn. Nitric oxide is the product of incomplete combustion, whereas a mixture of nitrogen dioxide and nitrogen tetroxide is the product of complete combustion.

TABLE 2.1
Nitrogen: Physical Properties

Chemical formula	N_2
Molecular weight	28.01
Density of gas @ 70°F	0.072 lb/ft³
Specific gravity of gas @ 70°F and 1 atm (air = 1)	0.967
Specific volume of gas @ 70°F and 1 atm	13.89 ft³
Boiling point @ 1 atm	−320.4°F
Melting point @ 1 atm	−345.8°F
Critical temperature	−232.4°F
Critical pressure	493 psia
Critical density	19.60 lb/ft³
Latent heat of vaporization @ boiling point	85.6 Btu/lb
Latent heat of fusion @ melting point	11.1 Btu/lb

The nitrogen oxides are usually collectively symbolized by the formula NO_x. USEPA, under the Clean Air Act (CAA), regulates the amount of nitrogen oxides that commercial and industrial facilities may emit to the atmosphere. The primary and secondary standards are the same: The annual concentration of nitrogen dioxide may not exceed 100 µg/m³ (0.05 ppm).

Much more will be said about primary and secondary air standards under CAA and nitrogen oxides later in the text.

ATMOSPHERIC OXYGEN

Oxygen (O_2; Greek *oxys*, "acid," and *genes*, "forming") constitutes approximately a fifth (21% by volume and 23.2% by weight) of the air in earth's atmosphere. Gaseous oxygen (O_2) is vital to life as we know it. On earth, oxygen is the most abundant element. Most oxygen on earth is not found in the free state, but in combination with other elements as chemical compounds. Water and carbon dioxide are common examples of compounds that contain oxygen, but there are countless others.

At ordinary temperatures, oxygen is a colorless, odorless, tasteless gas that supports not only life but also combustion. All the elements except the inert gases combine directly with oxygen to form oxides. However, oxidation of different elements occurs over a wide range of temperatures.

Oxygen is nonflammable but it readily supports combustion. All materials that are flammable in air burn much more vigorously in oxygen. Some combustibles, such as oil and grease, burn with nearly explosive violence in oxygen if ignited.

Physical Properties

The physical properties of oxygen are noted in Table 2.2.

Uses

The major uses of oxygen stem from its life-sustaining and combustion-supporting properties. It also has many industrial applications (when used with other fuel gases such as acetylene), including metal cutting, welding, hardening, and scarfing.

TABLE 2.2
Oxygen: Physical Properties

Chemical formula	O_2
Molecular weight	31.9988
Freezing point	−361.12°F
Boiling point	−297.33°F
Heat of fusion	5.95 Btu/lb
Heat of vaporization	91.70 Btu/lb
Density of gas @ boiling point	0.268 lb/ft³
Density of gas @ room temperature	0.081 lb/ft³
Vapor density (air = 1)	1.105
Liquid-to-gas expansion ratio	875

Ozone: Just Another Form of Oxygen

Ozone (O_3) is a highly reactive pale-blue gas with a penetrating odor. Ozone is an allotropic modification of oxygen. An allotrope is a variation of an element that possesses a set of physical and chemical properties significantly different from the normal form of the element. Only a few elements have allotropic forms; oxygen, phosphorous, and sulfur are some of them. Ozone is just another form of oxygen. It is formed when the molecule of the stable form of oxygen (O_2) is split by ultraviolet (UV) radiation or electrical discharge; it has three instead of two atoms of oxygen per molecule. Thus, its chemical formula is represented by O_3.

Ozone forms a thin layer in the upper atmosphere, which protects life on earth from ultraviolet rays, a cause of skin cancer. At lower atmospheric levels it is an air pollutant and contributes to the greenhouse effect. At ground level, ozone, when inhaled, can cause asthma attacks, stunted growth in plants, and corrosion of certain materials. It is produced by the action of sunlight on air pollutants, including car exhaust fumes, and is a major air pollutant in hot summers. More will be said about ozone and the greenhouse effect later in the text.

ATMOSPHERIC CARBON DIOXIDE

Carbon dioxide (CO_2) is a colorless, odorless gas (although it is felt by some persons to have a slight pungent odor and biting taste), is slightly soluble in water and denser than air (one and half times heavier than air), and is slightly acidic. Carbon dioxide gas is relatively nonreactive and nontoxic. It will not burn, and it will not support combustion or life.

CO_2 is normally present in atmospheric air at about 0.035% by volume and cycles through the biosphere (carbon cycle) as shown in Figure 2.2. Carbon dioxide, along with water vapor, is primarily responsible for the absorption of infrared energy reemitted by the earth, and in turn, some of this energy is reradiated back to

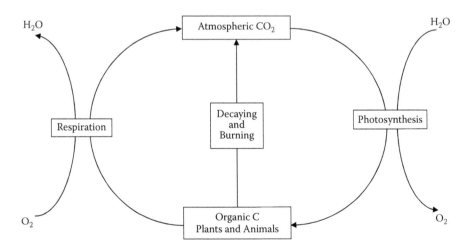

FIGURE 2.2 Carbon cycle.

the earth's surface. It is also a normal end product of human and animal metabolism. The exhaled breath contains up to 5.6% carbon dioxide. In addition, the burning of carbon-laden fossil fuels releases carbon dioxide into the atmosphere. Much of this carbon dioxide is absorbed by ocean water, some of it is taken up by vegetation through photosynthesis in the carbon cycle (see Figure 2.2), and some remains in the atmosphere. Today, it is estimated that the concentration of carbon dioxide in the atmosphere is approximately 350 parts per million (ppm) and is rising at a rate of approximately 20 ppm every decade. The increasing rate of combustion of coal and oil has been primarily responsible for this occurrence, which (as we will see later in this text) may eventually have an impact on global climate.

Physical Properties

The physical properties of carbon dioxide are noted in Table 2.3.

Uses

Solid carbon dioxide is used quite extensively to refrigerate perishable foods while in transit. It is also used as a cooling agent in many industrial processes, such as grinding, rubber work, cold-treating metals, vacuum cold traps, and so on.

Gaseous carbon dioxide is used to carbonate soft drinks, for pH control in water treatment, in chemical processing, as a food preservative, and in pneumatic devices.

ATMOSPHERIC ARGON

Argon (Ar; Greek *argos*, "idle") is a colorless, odorless, tasteless, nontoxic, nonflammable gaseous element (noble gas). It constitutes almost 1% of the earth's atmosphere and is plentiful compared to the other rare atmospheric gases. It is extremely inert and forms no known chemical compounds. It is slightly soluble in water.

Physical Properties

The physical properties of argon are noted in Table 2.4.

TABLE 2.3
Carbon Dioxide: Physical Properties

Chemical formula	CO_2
Molecular weight	44.01
Vapor pressure @ 70°F	838 psig
Density of the gas @ 70°F and 1 atm	0.1144lb/ft³
Specific gravity of the gas @ 70°F and 1 atm (air = 1)	1.522
Specific volume of the gas @ 70°F and 1 atm	8.741 ft³/lb
Critical temperature	−109.3°F
Critical pressure	1,070.6 psia
Critical density	29.2 lb/ft³
Latent heat of vaporization @ 32°F	100.8 Btu/lb
Latent heat of fusion @ −69.9°F	85.6 Btu/lb

TABLE 2.4
Argon: Physical Properties

Chemical formula	Ar
Molecular weight	39.95
Density of the gas @ 70°F	0.103 lb/ft³
Specific gravity of the gas @ 70°F	1.38
Specific volume of the gas @ 70°F	9.71 ft³/lb
Boiling point at 1 atm	−302.6°F
Melting point at 1 atm	−308.6°F
Critical temperature	−188.1°F
Critical pressure	711.5 psia
Critical density	33.444 lb/ft³
Latent heat of vaporization @ boiling point and 1 atm	69.8 Btu/lb
Latent heat of fusion	12.8 Btu/lb

Uses

Argon is used extensively in filling incandescent and fluorescent lamps and electronic tubes, to provide a protective shield for growing silicon and germanium crystals, and as a blanket in the production of titanium, zirconium, and other reactive metals.

ATMOSPHERIC NEON

Neon (Ne; Greek *neon*, "new") is a colorless, odorless, gaseous, nontoxic, chemically inert element. Air is about 2 ppt neon by volume.

Physical Properties

The physical properties of neon are noted in Table 2.5.

TABLE 2.5
Neon: Physical Properties

Chemical formula	Ne
Molecular weight	20.183
Density of the gas @ 70°F and 1 atm	0.05215 lb/ft³
Specific gravity of the gas @ 70°F and 1 atm	0.696
Specific volume of the gas @ 70°F and 1 atm	19.18 ft³/lb
Boiling point at 1 atm	−410.9°F
Melting point at 1 atm	−415.6°F
Critical temperature	−379.8°F
Critical pressure	384.9 psia
Critical density	30.15 lb/ft³
Latent heat of vaporization @ boiling point	37.08 Btu/lb
Latent heat of fusion	7.14 Btu/lb

Uses

Neon is used principally to fill lamp bulbs and tubes. The electronics industry uses neon singly or in mixtures with other gases in many types of gas-filled electron tubes.

ATMOSPHERIC HELIUM

Helium (He; Greek *helios*, "sun") is inert (and as a result, does not appear to have any major effect on, or role in, the atmosphere), nontoxic, odorless, tasteless, nonreactive, and colorless, forms no compounds, and makes about 0.00005% (5 ppm) by volume of air in the earth's atmosphere. Helium, as with neon, krypton, hydrogen, and xenon, is a noble gas. Helium is the second lightest element; only hydrogen is lighter. It is one-seventh as heavy as air. Helium is nonflammable and is only slightly soluble in water.

Physical Properties

The physical properties of helium are noted in Table 2.6.

ATMOSPHERIC KRYPTON

Krypton (Kr; Greek *kryptos*, "hidden") is a colorless, odorless, inert gaseous component of earth's atmosphere. It is present in very small quantities in the air (about 114 ppm).

Physical Properties

The physical properties of krypton are noted in Table 2.7.

Uses

Krypton is used principally to fill lamp bulbs and tubes. The electronics industry uses it singly or in mixture in many types of gas-filled electron tubes.

TABLE 2.6
Helium: Physical Properties

Chemical formula	He
Molecular weight	4.00
Density of the gas @ 70°F and 1 atm	0.0103 lb/ft³
Specific gravity of the gas @ 70°F and 1 atm	0.138
Specific volume of the gas @ 70°F and 1 atm	97.09 ft³/lb
Boiling point @ 1 atm	−452.1°F
Critical temperature	−450.3°F
Critical pressure	33.0 psia
Critical density	4.347 lb/ft³
Latent heat of vaporization @ boiling point and 1 atm	8.72 Btu/lb

TABLE 2.7
Krypton: Physical Properties

Chemical formula	Kr
Molecular weight	83.80
Density of the gas @ 70°F and 1 atm	0.2172 lb/ft^3
Specific gravity of the gas @ 70°F and 1 atm	2.899
Specific volume of the gas @ 70°F and 1 atm	4.604 ft^3/lb
Boiling point @ 1 atm	−244.0°F
Melting point @ 1 atm	−251°F
Critical temperature	−82.8°F
Critical pressure	798.0 psia
Critical density	56.7 lb/ft^3
Latent heat of vaporization @ boiling point	46.2 Btu/lb
Latent heat of fusion	8.41 Btu/lb

ATMOSPHERIC XENON

Xenon (Xe; Greek *xenon*, "stranger") is a colorless, odorless, nontoxic, inert, heavy gas that is present in very small quantities in the air (about 1 part in 20 million).

Physical Properties

The physical properties of xenon are noted in Table 2.8.

Uses

Xenon is used principally to fill lamp bulbs and tubes. The electronics industry uses it singly or in mixtures in many types of gas-filled electron tubes.

TABLE 2.8
Xenon: Physical Properties

Chemical formula	Xe
Molecular weight	131.3
Density of the gas @ 70°F and 1 atm	0.3416 lb/ft^3
Specific gravity of the gas @ 70°F and 1 atm	4.560
Specific volume of the gas @ 70°F and 1 atm	2.927 ft^3/lb
Boiling point at 1 atm	−162.6°F
Melting point at 1 atm	−168°F
Critical temperature	61.9°F
Critical pressure	847.0 psia
Critical density	68.67 lb/ft^3
Latent heat of vaporization at boiling point	41.4 Btu/lb
Latent heat of fusion	7.57 Btu/lb

ATMOSPHERIC HYDROGEN

Hydrogen (H$_2$; Greek *hydros* and *gen*, "water generator") is a colorless, odorless, tasteless, nontoxic, flammable gas. It is the lightest of all the elements and occurs on earth chiefly in combination with oxygen as water. Hydrogen is the most abundant element in the universe, where it accounts for 93% of the total number of atoms and 76% of the total mass. It is the lightest gas known, with a density approximately 0.07 that of air. Hydrogen is present in the atmosphere, occurring in concentrations of only about 0.5 ppm by volume at lower altitudes.

Physical Properties

The physical properties of hydrogen are noted in Table 2.9.

Uses

Hydrogen is used by refineries and petrochemical and bulk chemical facilities for hydrotreating, catalytic reforming, and hydrocracking. Hydrogen is used in the production of a wide variety of chemicals. Metallurgical companies use hydrogen in the production of their products. Glass manufacturers use hydrogen as a protective atmosphere in a process whereby molten glass is floated on a surface of molten tin. Food companies hydrogenate fats, oils, and fatty acids to control various physical and chemical properties. Electronics manufacturers use hydrogen at several steps in the complex processes for manufacturing semiconductors.

ATMOSPHERIC WATER

Leonardo da Vinci understood the importance of water when he said: "Water is the driver of nature." da Vinci was actually acknowledging what most scientists and many of the rest of us have come to realize: water, propelled by the varying

TABLE 2.9
Hydrogen: Physical Properties

Chemical formula	H$_2$
Molecular weight	2.016
Density of the gas @ 70°F and 1 atm	0.00521 lb/ft^3
Specific gravity of the gas @ 70°F and 1 atm	0.06960
Specific volume of the gas @ 70°F and 1 atm	192.0 ft^3/lb
Boiling point @ 1 atm	−423.0°F
Melting point @ 1 atm	−434.55°F
Critical temperature	−399.93°F
Critical pressure	190.8 psia
Critical density	1.88 lb/ft^3
Latent heat of vaporization @ boiling point	191.7 Btu/lb
Latent heat of fusion	24.97 Btu/lb

temperatures and pressures in earth's atmosphere, allows life as we know it to exist on our planet (Graedel and Crutzen, 1995).

The water vapor content of the lower atmosphere (troposphere) is normally with a range of 1–3% by volume, with a global average of about 1%. However, the percentage of water in the atmosphere can vary from as little as 0.1% to as much as 5% water, depending upon altitude; water in the atmosphere decreases with increasing altitude. Water circulates in the atmosphere in the hydrologic cycle, as shown in Figure 2.3.

Water vapor contained in earth's atmosphere plays several important roles: (1) it absorbs infrared radiation; (2) it acts as a blanket at night, retaining heat from the earth's surface; and (3) it affects the formation of clouds in the atmosphere.

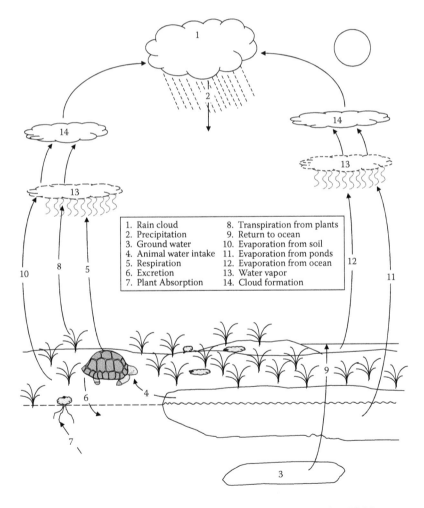

1. Rain cloud
2. Precipitation
3. Ground water
4. Animal water intake
5. Respiration
6. Excretion
7. Plant Absorption
8. Transpiration from plants
9. Return to ocean
10. Evaporation from soil
11. Evaporation from ponds
12. Evaporation from ocean
13. Water vapor
14. Cloud formation

FIGURE 2.3 Water cycle. Modified from Carolina Biological Supply Co. (1966).

ATMOSPHERIC PARTICULATE MATTER

There are significant numbers of particles (particulate matter) suspended in the atmosphere, particularly the troposphere. These particles originate in nature from smokes, sea sprays, dusts, and the evaporation of organic materials from vegetation. There is also a wide variety of nature's living or semiliving particles—spores and pollen grains, mites and other tiny insects, spider webs, and diatoms. The atmosphere also contains a bewildering variety of anthropogenic (man-made) particles produced by automobiles, refineries, production mills, and many other human activities.

Atmospheric particulate matter varies greatly in size (colloidal-sized particles in the atmosphere are called aerosols—usually less than 0.1 μm in diameter). The smallest are gaseous clusters and ions and submicroscopic liquids and solids. Somewhat larger ones produce the beautiful blue haze in distant vistas; those two to three times larger are highly effective in scattering light; and the largest consist of such things as rock fragments, salt crystals, and ashy residues from volcanoes, forest fires, or incinerators.

The numbers of which particulates are concentrated in the atmosphere vary greatly—ranging from more than 10,000,000/cubic centimeter (cc) to less than 1/L (0.001/cc). Excluding the particles in gases as well as vegetative material, sizes range from 0.005 to 500 microns, a variation in diameter of 100,000 times.

The largest number of airborne particulates is always in the invisible range. These numbers vary from less than 1 liter to more than a half million per cubic centimeter in heavily polluted air, and to at least ten times more than that when a gas-to-particle reaction is occurring (Schaefer and Day, 1981).

Based on particulate level, there are two distinct regions in the atmosphere: very clean and dirty. In the clean parts there are so few particulates that they are almost invisible, making them hard to collect or measure. In the dirty parts of the atmosphere—the air of a large metropolitan area—the concentration of particles includes an incredible variety of particulates from a wide variety of sources.

Atmospheric particulate matter performs a number of functions and undergoes several processes, and is involved in many chemical reactions in the atmosphere. Probably the most important function of particulate matter in the atmosphere is their action as nuclei for the formation of water droplets and ice crystals. Much of the work of Vincent J. Schaefer (inventor of cloud seeding) involved using dry ice in early attempts, but later evolved around the addition of condensing particles to atmospheres supersaturated with water vapor and the use of silver iodide, which forms huge numbers of very small particles. Another important function of atmospheric particulate matter is that they help determine the heat balance of the earth's atmosphere by reflecting light. Particulate matter is also involved in many chemical reactions in the atmosphere, such as neutralization, catalytic effects, and oxidation reactions. These chemical reactions will be discussed in greater detail later.

AIR FOR COMBUSTION

It is difficult to imagine where humans would be today or how far we would have progressed from our beginning to the present if we had not discovered and developed the use of fire. Today, of course, we are quite familiar with fire. We use the terms *fire, combustion, oxidation,* and *burning* pretty much in the same light to mean one and the same thing. However, in regards to combustion and oxidation, there is a subtle difference between them. During combustion, two or more substances chemically unite. In practice, one of them is almost always atmospheric oxygen, but combustion reactions are known in which oxygen is not one of the reactants. Thus, it is more correct to describe combustion as a rapid oxidation—or fire.

To state that atmospheric air plays an important role in combustion is to understate its significance; that is, we are stating the obvious. Though air is important in combustion, it is the actual chemical reaction involved with combustion that most of us give little thought to.

Combustion is a chemical reaction—one in which a fuel combines with air (oxygen) with the evolution of heat: burning. The combustion of fuels containing carbon and hydrogen is said to be complete when these two elements are oxidized to carbon dioxide and water (e.g., the combustion of carbon: $C + O_2 = CO_2$). In air pollution control, it is incomplete combustion that concerns us. Incomplete combustion may lead to (1) appreciable amounts of carbon remaining in the ash, (2) emission of some of the carbon as carbon monoxide, and (3) reaction of the fuel molecules to give a range of products that are emitted as smoke.

AIR FOR POWER

Most industrial processes use gases to power systems of one type or another. The work is actually performed by a gas under pressure in the system. A gas power system may function as part of a process, such as heating and cooling, or it may be used as a secondary service system, such as compressed air. Compressed air is the gas most often found in industrial applications, but nitrogen and carbon dioxide are also commonly used. A system that uses a gas for transmitting force is called a pneumatic system. The word *pneumatic* is derived from the Greek word for an unseen gas. Originally, pneumatic referred only to the flow of air. Now it includes the flow of any gas in a system under pressure.

Pneumatic systems perform work in many ways, including operating pneumatic tools, door openers, linear motion devices, and rotary motion devices. Have you ever watched (heard) an automobile mechanic remove and replace a tire on your car? The device he or she uses to take off and put on tire lug nuts is a pneumatic (air-operated) wrench. Pneumatic hoisting equipment may be found in heavy fabricating environments, and pneumatic conveyors are used in the processing of raw materials. Pneumatic systems are also used to control flow valves in chemical process equipment and large air conditioning systems.

The pneumatic system in an industrial plant usually handles compressed air. As pointed out earlier, compressed air is used for operating portable air tools, such as drills, wrenches, and chipping tools; for vises, chucks, and other clamping devices; for movable locating stops; for operating plastic molding machines; and for supplying air used in manufacturing processes. Although the pieces of pneumatic equipment just described are different from each other, they all convert compressed air into work. Later we review some of the laws of force and motion and their relation to pneumatic principles.

REFERENCES AND RECOMMENDED READING

Compressed Gas Association, Inc. 1990. *Handbook of compressed gases.* 3rd ed. New York: Van Nostrand Reinhold.

Graedel, T. E., and Crutzen, P. J. 1995. *Atmosphere, climate, and change.* New York: Scientific American Library.

Schaefer, V. J., and Day, J. A. 1981. *Atmosphere: Clouds, rain, snow, storms.* Boston: Houghton Mifflin Company.

Spellman, F. R., and Whiting, N. 2006. *Environmental science and technology: Concepts and applications.* Rockville, MD: Government Institutes.

3 Gas Physics

The difference between science and the fuzzy subjects is that science requires reasoning, while those other subjects merely require scholarship.

R. A. Heinlein, 1973, p. 348

INTRODUCTION

It was pointed out earlier that the air science practitioner must be well grounded in science fundamentals. Although the environmental engineer usually designs air pollution control equipment, if you are or will be responsible for the proper operation and selection of such equipment, it is incumbent upon you to know as much as possible about air, the ingredients that make up air, and the flow of air.

A knowledge of air physics (or more correctly stated, physics of gases and particles) is foundational to gaining an understanding of the characteristics of a system in which control equipment is used. For example, if you are required to select most air handling equipment (including fans, simple duct work, and collection equipment), you must be able to determine the volume of air to be handled.

In addition, you will need to have a good working knowledge of a few basic physical properties, such as gas density, pressure drop, viscosity of the gas, and pressure drop in filter media. This chapter discusses these basic physical properties, which are critical to gaining a better understanding of equipment operation. Because their application is important to air quality and emissions sampling/monitoring, emission assessment procedures, data summarization, engineering controls, and air quality monitoring, special attention is given to the gas laws.

BASIC MATH REVIEW

Over the years, countless studies have been conducted in an attempt to determine why many American students avoid science in school or avoid any branch of science as their chosen vocation. After years of study, several different opinions, and countless numbers of dollars spent in trying to determine why it is that students avoid majoring in scientific disciplines (including the science of air), it is amazing that most of these did not go to the source to find the answers—the students, of course.

In having done just that at Old Dominion University, I have found that most students avoid science or any thought of pursuing a vocation involving the sciences for two reasons: (1) many science curriculums that lead to science-related vocations require completion of several hours of study in foreign languages (primarily French), and (2) all scientific disciplines require an emphasis in mathematics. Many students I have surveyed and spoken to cannot see the efficacy or need to learn a foreign language. Some colleges require foreign language study with any 4-year program; others do not.

Regarding the second requirement—a need for a strong background in mathematics to succeed in science—students have no way around or out of this necessity. To work with, around, or even at the periphery of science, you must have a strong background in mathematics. So why are students so fearful of math? For the many students I have found that dislike or fear math, it is because it is difficult, time consuming, and requires a lot of work and absolute precision. In regards to math being difficult, maybe there are other factors involved. As one math teacher put it: "Those who have difficulty in math often do not lack the ability for mathematical calculation, they merely have not learned, or have not been taught, the 'language of math.'" (Price, 1991). Price's point is well taken and it can be expanded to: "The language of mathematics is a universal language." Mathematical symbols have the same meaning to people speaking in many different languages throughout the world.

What the problem with mathematics boils down to is that students do not understand the language of mathematics. They are not familiar with the symbols, definitions, and terms of mathematics. You cannot take shortcuts in this important subject, and some effort is required. Just as with any other subject, mathematics comes easily to some people and is difficult for others. Honest effort is all that is required to learn the language of mathematics.

What is mathematics? Mathematics is numbers. Math uses combinations of numbers and symbols to solve practical problems. Most of us, even the "mathophobes," handle what we have to in order to go through everyday life—we pay the bills and balance the checkbook. Advanced scientific endeavors require a more serious exploration of higher mathematics, but for environmental science or health at this level, regular garden variety high school mathematics will fit the bill. Since we all use numbers every day, then we are all mathematicians, to a point.

In science, we must take math beyond "to a point." We need to learn, understand, and appreciate mathematics. But how do we do this without failing? Probably the greatest single cause of failure to understand and appreciate mathematics is not knowing the key definitions of the terms used. In mathematics, more than in any other subject, each word used has a definite and fixed meaning.

The following basic definitions should be memorized. They will aid you in learning the material that follows.

Note: Readers who are current and well versed in the language of basic mathematics may want to skip over the following math review to the section titled "Plane Geometry." Many of the concepts and much of the material presented below are from the EPA's "Basic Concepts in Environmental Sciences" (2007).

MATH DEFINITIONS

An *integer* (or an *integral number*) is a whole number; 1, 2, 3, 4, 5, 6, 7, 8, 9, 10, 11, and 12 are the first twelve positive integers.

A *factor* (or *divisor*) of a whole number is any other whole number that exactly divides it. Thus, 2 and 5 are factors of 10. A *prime number* in math is a number that has no factors except itself and 1. Examples of prime numbers are 1, 3, 5, 7, and 11.

A *composite number* is a number that has factors other than itself and 1. Examples of composite numbers are 4, 6, 8, 9, and 12.

A *common factor* (or *common divisor*), or two or more numbers, is a factor that will exactly divide each of them. If this factor is the largest one possible, it is called the *greatest common divisor*. Thus, 3 is a common divisor of 9 and 27, but 9 is the greatest common divisor of 9 and 27.

A *multiple* of a given number is a number that is exactly divisible by the given number. If a number is exactly divisible by two or more other numbers, it is a common multiple of them. The least (smallest) such number is called the *lowest common multiple*. Thus, 36 and 72 are common multiples of 12, 9, and 4; however, 36 is the lowest common multiple.

A *product* is the result of multiplying two or more numbers together. Thus, 25 is the product of 5 × 5. Also, 4 and 5 are factors of 20.

A *quotient* is the result of dividing one number by another. For example, 5 is the quotient of 20 divided by 4.

A *dividend* is a number to be divided; a *divisor* is a number that divides. For example, in 100 ÷ 20 = 5, 100 is the dividend, 20 is the divisor, and 5 is the quotient.

We often have to apply mathematics in solving a problem. Sometimes the logic of the rule is apparent, but often it is not. Rules whose logic is not immediately obvious may be the result of experience, experiment, or merely rules of thumb. We cannot solve mathematical problems without knowing the key definitions and without following the rules.

Since the study of air is a science, mathematics plays an important role in solving problems related to air problems. As mentioned, the mathematics review presented in the following sections is very basic. For those seeking advanced studies of air, higher mathematics is required. However, in this text, mathematics beyond basic algebra is not required.

UNITS OF MEASUREMENT

A fundamental knowledge of units of measurement and how to use them is essential for students of air science. Air science students and practitioners should be familiar with both the U.S. Customary System (USCS) or English System and the International System of Units (SI). Some of the important units are summarized here to enable better understanding of material covered later in the text. Table 3.1 gives conversion factors between SI and USCS for some of the most basic units we will encounter.

In the study of air science, encountering both extremely large quantities and extremely small ones is quite common. The concentration of some toxic substance may be measured in parts per million or billion (ppm or ppb). (For example, ppm may be roughly described as an amount contained in a shot glass in the bottom of a swimming pool.) To describe such large or small quantities, a system of prefixes that accompany the units is useful. Some of the more important prefixes are presented in Table 3.2.

UNITS OF MASS

Simply defined, mass is a quantity of matter and measurement of the amount of inertia that a body possesses. Mass expresses the degree to which an object resists

TABLE 3.1
Commonly Used Units and Conversion Factors

Quantity	SI Units	SI Symbol ×	Conversion Factor =	USCS Units
Length	Meter	m	3.2808	ft
Mass	Kilogram	kg	2.2046	lb
Temperature	Celsius	°C	1.8(°C) + 32	°F
Area	Square meter	m^2	10.7639	ft^2
Volume	Cubic meter	m^3	35.3147	ft^3
Energy	Kilojoule	kJ	0.9478	Btu
Power	Watt	W	3.4121	Btu/h
Velocity	Meter/second	m/s	2.2369	mi/h

a change in its state of rest or motion and is proportional to the amount of matter in the object.

Beginning science students often confuse mass with weight, but they are different. *Weight* is the gravitational force action upon an object and is proportional to mass. In the SI system (a modernized metric system), the fundamental unit of mass is 1 gram (g). To show the relationship between mass and weight, consider that there is 452.6 g per pound. In laboratory-scale operations, the gram is a convenient unit of measurement. However, in real-world applications the gram is usually prefixed with one of the prefixes shown in Table 3.2. For example, human body mass is expressed in kilograms (1 kg = 2.2 pounds, which is the mass of 1 liter of water). When dealing with environmental conditions such as air pollutants and toxic water pollutants, we may measure in teragrams (1×10^{12} g) and micrograms (1×10^{-6} g), respectively. When dealing with large-scale industrial commodities, we may measure mass in units of megagrams (Mg), which is also known as a metric ton.

TABLE 3.2
Common Prefixes

Quantity	Prefix	Symbol
10^{-12}	pico	p
10^{-9}	nano	n
10^{-6}	micro	μ
10^{-3}	milli	m
10^{-2}	centi	c
10^{-1}	deci	d
10	deca	da
10^2	hecto	h
10^3	kilo	k
10^6	mega	M

Often, mass and density are mistakenly thought of as signifying the same thing; they do not. Where mass is the quantity of matter and measurement of the amount of inertia that a body contains, *density* refers to how compacted a substance is with matter. Density is the mass per unit volume of an object, and its formula can be written as

$$\text{Density} = \frac{\text{Mass}}{\text{Volume}} \tag{3.1}$$

Thus, something with a mass of 25 kg that occupies a volume of 5 m³ would have a density of 25 kg/5 m³ = 5 kg/m³.

UNITS OF LENGTH

In measuring locations and sizes, we use the fundamental property of *length*, which is defined as the measurement of space in any direction. Space has three dimensions, each of which can be measured by a length. This can be easily seen by considering the rectangular object shown in Figure 3.1. It has length, width, and height, but each of these dimensions is a length.

In the metric system, length is expressed in units based on the *meter* (m), which is 39.37 inches long. A kilometer (km) is equal to 1,000 m, and is used to measure relatively great distances. In practical laboratory applications, the centimeter (cm = 0.01 m) is often used. There are 2.540 cm per inch, and the centimeter is employed to express lengths that would be given in inches in the English system. The micrometer (μm) is also commonly used to express measurements of bacterial cells and wavelengths of infrared radiation by which Earth reradiates solar energy back to outer space. For measuring visible light (400 to 800 nm), the nanometer (nm) (10^{-9}) is often used.

UNITS OF VOLUME

The easiest way in which to approach measurements involving volume is to remember that volume is surface area times a third dimension. The *liter* is the basic metric unit of volume and is the volume of a decimeter cubed (1 L = 1 dm³). A milliliter (ml) is the same volume as a cubic centimeter (cm³).

UNITS OF TEMPERATURE

Temperature is a measure of how hot something is or how much thermal energy it contains. Temperature is a fundamental measurement in air science, especially

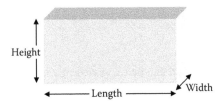

FIGURE 3.1 Length

in most pollution work. The temperature of a stack gas plume, for example, determines its buoyancy and how far the plume of effluent will rise before attaining the temperature of its surroundings. This, in turn, determines how much it will be diluted before traces of the pollutant reach ground level.

Temperature is measured on several scales; for example, the *centigrade* (*Celsius*) and *Fahrenheit* scales are both measured from a reference point—the freezing point of water—which is 0°C or 32°F. The boiling point of water is 100°C or 212°F. For thermodynamic devices, it is usual to work in terms of absolute or thermodynamic temperature where the reference point is absolute zero, which is the lowest possible temperature attainable. For absolute temperature measurement, the thermodynamic unit, or *Kelvin* (K) scale, is used, which uses centigrade divisions for which zero is the lowest attainable measurement. A unit of temperature on this scale is equal to a Celsius degree but is not called a degree; it is called a Kelvin and is designated K, not °K. The value of absolute zero on the Kelvin scale is –273.15°C, so that the Kelvin temperature is always a number 273 (rounded) higher than the Celsius temperature. Thus, water boils at 373 K and freezes at 273 K.

It is easy to convert from the Celsius scale to the Kelvin scale. Simply add 273 to the Celsius temperature and you have the Kelvin temperature. Mathematically,

$$K = C + 273$$

where
K = temperature on the Kelvin scale
C = temperature on the Celsius scale

Converting from Fahrenheit to Celsius or vice versa is not so easy. The equations used are

$$C = 5/9(F - 32)$$

and

$$F = 9/5C + 32$$

where
C = temperature on the Celsius scale
F = temperature on the Fahrenheit scale

As examples, 15°C = 59°F and 68°F = 20°C. F and C, of course, can be negative numbers.

UNITS OF PRESSURE

Pressure is force per unit area and can be expressed in a number of different units, including the *atmosphere* (atm), which is the average pressure exerted by air at sea level, or the *pascal* (Pa), usually expressed in kilopascal (1 kPa = 1,000 Pa, and 101.3 kPa = 1 atm). Pressure can also be given as *millimeters of mercury* (mmHg), which is based on pressure required to hold up a column of mercury in a mercury barometer. One millimeter of mercury is a unit called the *torr*, and 760 torr = 1 atm.

UNITS OFTEN USED IN AIR STUDIES

In air studies, it is often the concentration of some substance (foreign or otherwise) in air that is of interest. In either the gaseous or liquid medium, concentrations may be based on volume or weight, or a combination of the two, which may lead to some confusion. To understand how weight and volume are used to determine concentrations when studying liquids or gases/vapors, the following explanations are provided.

LIQUIDS

Concentrations of substances dissolved in water are usually expressed in terms of weight of substance per unit volume of mixture. In environmental science, a good practical example of this weight per unit volume is best observed whenever a contaminant is dispersed in the atmosphere in solid or liquid form as a mist, dust, or fume. When this occurs, its concentration is usually expressed on a weight-per-volume basis. Outdoor air contaminants and stack effluents are frequently expressed as grams, milligrams, or micrograms per cubic meter; ounces per thousand cubic feet; pounds per thousand pounds of air; and grains per cubic foot. Most measurements are expressed in metric units. However, the use of standard U.S. units is justified for purposes of comparison with existing data, especially those relative to the specifications for air-moving equipment.

Alternatively, concentrations in liquids are expressed as weight of substance per weight of mixture, with the most common units being parts per million (ppm) or parts per billion (ppb). Since most of the concentrations of pollutants are very small, 1 liter of mixture weighs essentially 1,000 grams, so that for all practical purposes we can write:

$$1 \text{ mg/L} = 1 \text{ g/m}^3 = 1 \text{ ppm (by weight)}$$
$$1 \text{ } \mu\text{g/L} = 1 \text{ mg/m}^3 = 1 \text{ ppb (by weight)}$$

The air science practitioner may also be involved with concentrations of liquid wastes (which may contaminate the atmosphere) that are so high that the *specific gravity* (the ratio of an object's or substance's weight to that of an equal volume of water) of the mixture is affected, in which case a correction to the above may be required:

$$\text{mg/L} = \text{ppm (by weight)} \times \text{specific gravity} \qquad (3.2)$$

GASES/VAPORS

For most air pollution work, it is customary to express pollutant concentrations in volumetric terms. For example, the concentration of a gaseous pollutant in parts per million (ppm) is the volume of pollutant per million volumes of the air mixture. That is,

$$ppm = \frac{\text{Parts of contaminant}}{\text{Million parts of air}} \qquad (3.3)$$

Note that calculations for gas and vapor concentrations are based on the gas laws:

- The volume of gas under constant temperature is inversely proportional to the pressure.
- The volume of a gas under constant pressure is directly proportional to the Kelvin temperature. The Kelvin temperature scale is based on absolute zero (0°C = 273 K).
- The pressure of a gas of a constant volume is directly proportional to the Kelvin temperature.

Thus, when measuring contaminant concentrations, you must know the atmospheric temperature and pressure under which the samples were taken. At *standard temperatures and pressure* (STP), 1 g-mol of an ideal gas occupies 22.4 L. The STP is 0°C and 760 mmhg. If the temperature is increased to 25°C (room temperature) and the pressure remains the same, 1 g-mol of gas occupies 24.45 L.

Sometimes it is necessary to convert milligrams per cubic meter (mg/m³)—weight-per-volume ratio—into a volume-per-unit volume ratio. If it is understood that 1 g-mole of an ideal gas at 25°C occupies 24.45 L, the following relationships can be calculated:

$$ppm = \frac{24.45}{\text{Molecular wt}} \; mg/m^3 \qquad (3.4)$$

$$mg/m^3 = \frac{\text{Molecular wt}}{24.45} \; ppm \qquad (3.5)$$

POWERS OF 10 AND SCIENTIFIC NOTATION

Note: In practice, the air science practitioner should realize that the accuracy of a final answer can never be better than the accuracy of the data used. Furthermore, remember that correct and accurate data are worthless unless the operator is able to make correct computations.

We describe two common methods of expressing a number in this section: powers of 10 and scientific notation.

POWERS OF 10 NOTATION

An expression such as 5^7 is a shorthand method of writing multiplication. For example, 5^7 can be written as

$$5 \times 5 \times 5 \times 5 \times 5 \times 5 \times 5$$

The expression 5^7 is referred to as 5 to the seventh power and is composed of an exponent and a base number. The exponent (or power of) indicates how many times a member is to be multiplied together. The base is the number being multiplied:

$$^7 \text{(exponent)}$$
$$5 \text{ (base)}$$

These same considerations apply to letters (a, b, x, y, etc.) as well. For example:

$$z^2 = (z)(z) \text{ or } z^4 = (z)(z)(z)(z)$$

When a number or letter does not have an exponent, it is considered to have an exponent of 1. Thus,

$$5 = 5^1 \text{ or } z = z^1$$

The following examples help to illustrate the concept of powers notation.

How is the term $(3/8)^2$ written in expanded form? When parentheses are used, the exponent refers to the entire term within the parentheses. Thus,

$$(3/8)^2 = (3/8)(3/8)$$

When a negative exponent is used with a number or term, a number can be reexpressed using a positive exponent:

$$6^{-3} = 1/6^3$$

Note: Any number or letter such as 3^0 or X^0 does not equal 3×1 or X1, but simply 1.

When a term is given in expanded form, you can determine how it would be written in exponential form. For example,

$$(5)(5)(5) = 5^3$$

or

$$(\text{in.})(\text{in.}) = \text{in.}^2$$

We commonly see powers used with a number or term to denote area or volume units such as square inches or cubic feet (in.^2, ft^2, in.^3, ft^3).

In moving a power from the numerator of a fraction to the denominator, or vice versa, the sign of the exponent is changed. For example,

$$\frac{3^3 \times 4^{-2}}{8} = \frac{3^3}{8 \times 4^2}$$

SCIENTIFIC NOTATION

Scientific notation is a method by which any number can be expressed as a term multiplied by a power of 10. The term is always greater than or equal to 1 but less than 10. Examples of powers of 10 are:

$$3.2 \times 10^1$$
$$1.8 \times 10^3$$
$$9.550 \times 10^4$$
$$5.31 \times 10^{-2}$$

The numbers can be taken out of scientific notation by performing the indicated multiplication. For example,

$$3.2 \times 10^1 = (3.2)(10) = 32$$
$$1.8 \times 10^3 = (1.8)(10)(10)(10) = 1,800$$
$$9.550 \times 10^4 = (9.550)(10)(10)(10)(10) = 95,500$$
$$5.31 \times 10^{-2} = (5.31)1/10^2 = 0.351$$

An easier way to take a number out of scientific notation is by moving the decimal point the number of places indicated by the exponent.

Rule 1

Multiply by the power 10 indicated. A positive exponent indicates a decimal move to the *right*, and a negative one indicates a decimal move to the *left*.

Using the same examples above, the decimal point move rather than the multiplication method is performed as follows:

$$3.2 \times 10^1$$

The positive exponent of 1 indicates that the decimal point in 3.2 should be moved one place to the right:

$$3.2 \times 32$$

The next example is

$$1.8 \times 10^3$$

The positive exponent of 3 indicates that the decimal point in 1.8 should be moved three places to the right:

$$1.800 \times 1,800$$

The next example is

$$9.550 \times 10^4$$

The positive exponent of 4 indicates that the decimal point should be moved four places to the right:

$$9.5500 = 95,500$$

The final example is

$$5.31 \times 10^{-2}$$

The negative exponent of 2 indicates that the decimal point should be moved two places to the left:

$$05.31 = 0.0531$$

There are very few instances in which you will need to put a number into scientific notation, but you should know how to do it, if required.

Procedure: When placing a number into scientific notation, place a decimal point after the first nonzero digit. (Remember that if no decimal point is shown in the number to be converted, it is assumed to be at the end of the number.) Count the number of places from the standard position to the original decimal point. This represents the exponent of the power of 10.

Rule 2

When a number is put into scientific notation, a decimal point move to the *left* indicates a positive exponent, and a decimal point move to the *right* indicates a negative exponent.

Now let us try converting a few numbers into scientific notation, e.g., 1,500. Remember, in order to obtain a number between 1 and 9, the decimal point must be moved three places to the left. The number of place moves (three) becomes the exponent of the power of 10, and the move to the left indicates a positive exponent:

$$1,500 = 1.5 \times 10^3$$

Let us try decimal number 0.0661:

$$0.0661 = 6.61 \times 10^{-2}$$

Two place moves to the right indicates a negative exponent of 2.

LOGARITHMS (EPA, 2007)

A logarithm is a way of expressing numbers as a function of a base number such as 10. For example, 2 is the logarithm of 100 in base 10. This logarithmic function and its equivalent exponential form are shown in the three examples below. It is apparent that the logarithmic function is the inverse of the exponential function.

Logarithmic Function	Exponential Function
$2 = \log_{10}(100)$	$10^2 = 100$
$5.16325 = \log_{10}(145,630)$	$10^{5.16325} = 145,630$
$6.41162 = \log_{10}(2,580,000)$	$10^{6.41162} = 2,580,000$

The logarithmic function provides a way to express a very large number concisely. For example, refrigeration or cryogenic condensers can be added to the air pollution control system to reduce gaseous emissions. As the condenser cools the gas stream, the temperature and vapor pressure of the gas decrease. The vapor pressure of organic compounds in the gas stream can vary from values as low as 0.001 mmHg to more than 100 mmHg, depending on the operating temperature in the condenser. Thus, the vapor pressure can span a 100,000 mmHg range. It is difficult to express the vapor pressure at low gas temperatures without the use of logarithms. Logarithms allow you to convert exponential data to a format that can easily fit on a graph. Using logarithms is also helpful in the field of air pollution when important operating conditions vary over a wide range. For example, the pH scale is a logarithmic way to express H^+ (sometimes written H_3O^+) ion concentration over a fourteen order of magnitude scale.

The laws of handling logarithms are similar to those for handling exponents. Some examples are illustrated below.

$$\text{Log}_{10}\, xy = \log_{10} x + \log_{10} y$$
$$\text{Log}_{10}\,(x/y) = \log_{10} x - \log_{10} y$$

NATURAL LOGARITHMS

In addition to the base of 10, the natural (sometimes termed Napierian) base is often used in air pollution–related work. The natural base is the number 2.718282. The number is often expressed as the letter e. For example, the expression e^2 is equivalent to $(2.718282)^2$. When natural logs are used, the symbol is written as Ln rather than $\text{Log}_{2.718282}$, as shown below.

Logarithmic Function	Exponential Function
Ln $(y) = x$	$e^x = y$

EQUATIONS: SOLVING FOR THE UNKNOWN

In environmental science applications related to air measurements and calculations, you may use equations to solve for the unknown quantity. To make these calculations, you must first know the values for all but one of the terms of the equation to be used.

An *equation* is a statement that two expressions or quantities are equal in value. The statement of equality $5x + 4 = 19$ is an equation; that is, it is algebraic shorthand for "The sum of five times a number plus 4 is equal to 19." It can be seen that the equation $5x + 4 = 19$ is much easier to work with than the equivalent sentence.

When thinking about equations, it is helpful to consider an equation as being similar to a balance. The equal sign tells you that two quantities are in balance (i.e., they are equal).

Let us get back to the equation $5x + 4 = 19$. The solution to this problem may be summarized in three steps:

$$(1)\ 5x + 4 = 19$$
$$(2)\ 5x = 15$$
$$(3)\ x = 3$$

Step 1 expresses the whole equation. In step 2, 4 has been subtracted from both members of the equation. In step 3, both members have been divided by 5.

An equation is therefore kept in balance (both sides of the equal sign are kept equal) by subtracting the same number from both members (sides), adding the same number to both, or dividing or multiplying by the same number.

The expression $5x + 4 = 19$ is called a *conditional equation* because it is true only when x has a certain value. The number to be found in a conditional equation is called the *unknown number, unknown quantity,* or more briefly, the *unknown.*

Solving an equation is finding the value(s) of the unknown that make the equation true.

Another equation, one that practitioners of air and science disciplines should be familiar with, is

$$W = F \times D \qquad\qquad (3.6)$$

where
  W = work
  F = force
  D = distance

Thus,

$$\text{Work} = \text{Force (pounds)} \times \text{Distance (feet or inches)}$$
$$= \text{Foot-pounds or inch-pounds}$$

To demonstrate an equation the air science practitioner may be called upon to use, consider the following situation. Fabric filters are commonly used to separate dry particles from a gas stream, usually of air or combustion gases. In fabric filtration, the particulate-laden gas flows into and through a number of filter bags placed in parallel, leaving the particulates retained by the fabric (more will be said about fabric filters later).

$$V = \frac{Q}{A}$$ (3.7)

where
  V = superficial filtering velocity (aka air/cloth ratio)
  Q = volumetric gas flow rate, m^3/min
  A = cloth area, m^2

The terms of these equations are W, F, D and V, Q, A. In solving problems using these equations, you would need to be given values to substitute for any two of the three terms in each equation. Again, the term for which you do not have information is called the unknown, which is often indicated by a letter such as x, y, or z, but may be any letter.

Suppose you have this equation:

$$80 = (x)(4)$$

How can you determine the value of x? By following the axioms presented below, the solution to the unknown is quite simple.

1. If equal numbers are added to equal numbers, the sums are equal.
2. If equal numbers are subtracted from equal numbers, the remainders are equal.
3. If equal numbers are multiplied by equal numbers, the products are equal.
4. If equal numbers are divided by equal numbers (except zero), the quotients are equal.
5. Numbers that are equal to the same number or to equal numbers are equal to each other.
6. Like powers of equal numbers are equal.
7. Like roots of equal numbers are equal.
8. The whole of anything equals the sum of all its parts.

Note: Axioms 2 and 4 were used to solve the equation $5x + 4 = 19$.

Here you can see by inspection that $x = 3$, but inspection does not help in solving more complicated equations. But if you notice that to determine $x = 3$, 4 is added to each member of the given equation, you have acquired a method or procedure that can be applied to similar but more complex problems.

Given equation

$$x - 6 = 3$$

add 6 to each member (axiom 1),

$$x = 3 + 6$$

Collecting the terms (that is, adding 3 and 6),

$$x = 9$$

After you have obtained a solution to an equation, you should always check it—an easy process. All you need do is substitute the solution for the unknown quantity in the given equation. If the two members of the equation are then identical, the number substituted is the correct answer.

EXAMPLE 3.1
Solve and check $4x + 5 - 7 = 2x + 6$

Solution

$$4x + 5 - 7 = 2x + 6$$
$$4x - 2 = 2x + 6$$
$$4x = 2x + 8$$
$$2x = 8$$
$$x = 4$$

Substituting the answer $x = 4$ in the original equation,

$$4x + 5 - 7 = 2x + 6$$
$$4(4) + 5 - 7 = 2(4) + 6$$
$$16 + 5 - 7 = 8 + 6$$
$$14 = 14$$

Because the statement $14 = 14$ is true, the answer $x = 4$ must be correct.

SETTING UP EQUATIONS

The equations discussed in the preceding paragraphs were expressed in *algebraic* language. You must learn how to set up an equation by translating a sentence into an equation (into algebraic language) and then solving this equation. The following suggestions and examples should help you:

1. Always read the statement of the problem carefully.
2. Select the unknown number and represent it by some letter. If more than one unknown quantity exists in the problem, try to represent those numbers in terms of the same letter—that is, in terms of one quantity.
3. Develop the equation, using the letter(s) selected, and then solve.

EXAMPLE 3.2

Five more than three times a number is the same as ten less than six times the number. What is the number?

Solution

Let n represent the number.

$$3n + 5 = 6n - 10$$
$$3n - 6n = -10 - 5$$
$$-3n = -15$$
$$n = 5$$

EXAMPLE 3.3

If five times the sum of a number and six is increased by three, the result is two less than ten times the number. Find the number.

Solution

Let n represent the number.

$$5(n + 6) + 3 = 10n - 2$$
$$5n + 33 = 10n - 2$$
$$5n = 10n - 35$$
$$n = 7$$

EXAMPLE 3.4

The greater of two numbers is three less than seven times the smaller. Also, twelve more than the greater is the same as ten times the smaller. Find both numbers.

$$7n - 3 + 12 = 10n$$
$$7n + 9 = 10n$$
$$7n - 10n = -9$$
$$-3n = -9$$
$$n = 3$$

The small number is 3.

$$7(n) - 3$$
$$7(3) - 3 = 18$$

The greater number is 18.

RATIO AND PROPORTION

RATIO

Ratio is the comparison of two numbers by division or an indicated division. The ratio of one number to another is determined when the one number is divided by the other.

A ratio always includes two numbers. For example, if a box has a length and width of 8 inches and 4 inches, the ratio of the length to the width is expressed as 8/4 or 8:4. Both expressions have the same meaning.

All ratios are reduced to the lowest possible terms, similar to reducing a fraction to the lowest possible terms. The ratio 8:4 should be reduced to its lowest possible terms by dividing the 8 and the 4 by 4. The resulting ratio is 2:1 or 2/1. The ratio of the length of the box to its width is 2:1, since the box is two times as long as it is wide.

This ratio can also be stated as the relationship of the width to the length. The box is 4 inches wide and 8 inches long. The ratio of the width to the length is 4:8 or 4/8. This ratio, when reduced, becomes 1:2 or 1/2. The width of the box is 1/2 its length.

Let us look at an example of how and where ratios are used in air science work. In the United States, concentrations of atmospheric gases and vapors are usually expressed as *mixing ratios* and reported in parts per million volume (ppmv). One ppmv is equal to a volume of a gas in 1 million volumes of air. (Note: Mixing ratios used to express air concentrations should not be confused with those used for water, which are weight/volume ratios (mg/L), and for solids, which are weight/weight ratios (μg/gm, mg/km). Although all are expressed as ppm, they are not equivalent concentrations.)

$$1 \text{ ppmv} = \frac{1 \text{ gas volume}}{10^6 \text{ air volumes}}$$

A microliter volume of gas mixed in a liter of air would therefore be equal to 1 ppmv.

$$1 \text{ ppmv} = \frac{1 \text{ μL gas}}{1 \text{ L air}}$$

Mixing ratios based on volume/volume ratios may also be expressed as parts per hundred million (pphmv), parts per billion (ppbv), or parts per trillion (pptv).

PROPORTIONS

Simply put, a *proportion* is a statement of equality between two ratios. Thus, 2:4 = 4:8 is a proportion. We know that the two ratios are equal when the proportion is written in fractional form: 2/4 = 4/8. Either form may be read as follows: "Two is to four as four is to eight."

A general statement of the preceding proportion would be:

$$\frac{a}{b} = \frac{c}{d} \text{ (fractional form)}$$

$$a:b = c:d \text{ (proportional form)}$$

where *a*, *b*, *c*, and *d* represent numbers.

A proportion may be written with a double colon (::) in place of the equal sign:

$$a:b::c:d$$

The first and last terms are called the extremes; the second and third terms are called the means.

When you consider proportions as fractions, it becomes evident that if any three members are known, then the fourth can be determined. When the proportion given earlier, $a{:}b = c{:}d$, is written as a fraction, $a/b = c/d$, then $a \times d = c \times b$. If you substitute numbers for letters and for the proportion, then $1/2 = 2/4$, $2 \times 2 = 4$, and $1 \times 4 = 4$. It is therefore obvious that the product of the means (2×2) equals the product of the extremes (1×4).

EXAMPLE 3.5
What is x in the proportion 2:3 = x:12?

Solution
Rewriting in fractional form,

$$\frac{2}{3} = \frac{x}{12}$$

Since $12 = 4 \times 3$, x must be 2×4, or 8, because this gives equal fractions. Another method for solving this proportion is use of the principle that the product of the means equals the product of the extremes. The product of the means, $3x$, equals the product of the extremes, 2×12, or 24. Thus, $3x = 24$, and x must equal 8, the same answer obtained earlier.

EXAMPLE 3.6
What is x in the proportion 5:x = 2,000:10,000

Solution
Rewrite in fractional form:

$$\frac{5}{x} = \frac{2,000}{10,000}$$

And solve for the unknown value

$$5 = \frac{(2,000)(x)}{10,000}$$

$$(5)(10,000) = (2,000)(x)$$

$$\frac{(5)(10,000)}{2,000} = x$$

$$25 = x$$

EXAMPLE 3.7
If a pump will fill a tank in 20 hours at 4 gpm (gallons per minute), how long will it take a 10 gpm pump to fill the same tank?

First, analyze the problem. Here the unknown is some number of hours. But should the answer be larger or smaller than 20 hours? If a 4 gpm pump can fill the tank in 20 hours, a larger pump (10 gpm) should be able to complete the filling in less than 20 hours. Therefore, the answer should be less than the 20 hours.

Now set up the proportion:

$$\frac{x \text{ hours}}{20 \text{ hours}} = \frac{4 \text{ gpm}}{10 \text{ gpm}}$$

$$x = \frac{(4)(20)}{10}$$

$$x = 8 \text{ hours}$$

It will not be long before you gain an understanding of proportion problems that will allow you to skip some of the various steps to solving these problems (practice makes perfect and repetition aids easy recognition). In the following examples, a shortcut method is shown that will allow an experienced operator to solve problems quite easily.

EXAMPLE 3.8

To make a certain chemical solution, 66.3 mg of chemical must be added to 150 L of water. How much of the chemical should be added to 25 L to make up the same strength solution?

To solve this problem, you must first decide what is unknown, and whether you expect the unknown value to be larger or smaller than the known value of the same unit. The amount of chemical to be added to 25 L is the unknown, and you would expect this to be smaller than the 66.3 mg needed for 150 L.

First, take the two known quantities of the same unit (25 L and 150 L) and make a fraction to multiply the third known quantity (66.3 mg) by. Notice that there are two possible fractions you can make with 25 and 150:

$$\frac{25}{150} \text{ or } \frac{150}{25}$$

Next, choose the fraction that will make the unknown number of milligrams less than the known. Multiplying 66.3 by the fraction 25/150 would result in a number smaller than 66.3.

$$\frac{25}{150}(66.3) = z$$

$$\frac{(25)(66.3)}{150} = z$$

$$11.05 \text{ mg} = z$$

From the above operation, it should be obvious that the key to this method is arranging the two known values of like units into a fraction that, when multiplied by the third known value, will render a result that is smaller or larger, as required.

EXAMPLE 3.9

If a machine metal is composed of 30 parts copper and 10 parts tin, what is the weight of each in a machine weighing 2,400 lb?

Solution

The total weight of the machine, in parts, is equal to $10 + 30$, or 40 parts. The ratio of the number of parts of each metal to the total number of parts of metal equals the ratio of the weight of each metal to the total weight of metal. To calculate the pounds of copper, set up the following ratio:

$$\frac{30}{40} = \frac{c}{2,400} \text{ or } 30 : 40 = c : 2,400$$

Then

$$30 \times 2,400 = c \times 40$$

$$c = \frac{30 \times 2,400}{40}$$

$$c = 1,800 \text{ lb}$$

For tin,

$$\frac{10}{40} = \frac{t}{2,400} \text{ or } 10 : 40 = t : 2,400$$

$$10 \times 2,400 = t \times 40$$

$$t = \frac{10 \times 2,400}{40}$$

$$t = 600 \text{ lb}$$

FINDING AVERAGES

Finding *averages* (or an arithmetic mean of a series of numbers) is accomplished by adding the numbers and dividing by the number of numbers in the group. This is an activity required on several computations that are made by air specialists (e.g., field studies, statistical analysis, etc.).

Averaging plays an important role in scientific analysis. Averaging allows you to group data (information) and then compute an average from which a trend may be determined. It is important to point out that an average is a reflection of the general nature of a certain group and does not necessarily reflect any one component of that group.

EXAMPLE 3.10

Find the average of the following series of numbers: 12, 14, 11, 6, 2, 9, 8, 7, 8, and 4. Adding the numbers together, we get 81. There are 10 numbers in this set, so we divide 81 by 10 to get 8.1 as the average of the set.

PERCENT

Simply put, *percent* (%) means "parts of 100 parts" or "by the hundred." Percent is used to describe portions of the whole. Thus, 12% means 12 percent or 12/100 or 0.12. As another example, consider a tank that is 6/10 full; we say that it contains 60% of the original solution or of its total capacity. Percent is also commonly used to describe the portion of a budget spent on a project completed, e.g., "There is only 15% of the budgeted amount remaining" or "The air sampling study is 40% complete."

Except when it is used in calculation, percentage is expressed as a whole number with a percent sign (%) after it. In a calculation, percent is expressed as a decimal. The decimal is obtained by dividing the percent by 100. For example, 12% is expressed as the decimal 0.12, since 12% is equal to 12/100. This decimal is obtained by dividing 12 by 100.

To determine what percentage a part is of the whole, divide the part by the whole. For example, if there are 110 sample blanks to label and Nancy has finished 40 of them, what percentage of the blanks has been labeled?

Step 1: $40 \div 110 = 0.36$.
Step 2: 0.36 is converted to percent by multiplying the answer by 100.
Step 3: $0.36 \times 100 = 36\%$. Thus, 36% of the 110 sample blanks have been labeled.

To determine the whole when the part and its percentage are given, divide the part by the percentage. Example: How much 65% calcium hypochlorite is required to obtain 15 pounds of chlorine? The part is 15 pounds, which is 65% of the whole.

Step 1: Convert the percentage to a decimal by dividing by 100:

$$65\% \div 100 = 0.65$$

Step 2: Divide the part by the decimal equivalent of the percentage:

$$15 \text{ lb} \div 0.65 = 23.1 \text{ (rounded)}$$

To increase a value by a percent, we add the decimal equivalent of the percent to 1 and multiply it times the number.

A filter bed will expand 20% during backwash. If the filter bed is 48 inches deep, how deep will it be during backwash?

Step 1: Change the percent to a decimal:

$$20\% \div 100 = 0.20$$

Step 2: Add the whole number 1 to this value:

$$1 + 0.20 = 1.20$$

Step 3: Multiply times the value:

$$48 \text{ inches} \times 1.20 = 57.6 \text{ inches}$$

In air science work, the concentration of chemicals used or detected is commonly expressed as a percentage. For the sake of simplicity, let us look at an example using liquid chemical solution.

Let us say we have a sulfur dioxide (an air pollutant in gaseous form) solution made to have a 6% concentration. It is often desirable to determine this concentration in mg/L. To accomplish this, we consider 6% as six percent of a million. (Note: A million because a liter of water weighs 1,000,000 mg, and 1 mg in 1 liter is 1 part in a million parts (ppm)). To find the concentration in mg/L when it is expressed in percent, do the following:

Step 1: Change the percent to a decimal:

$$6\% \div 100 = 0.06$$

Step 2: Multiply times a million:

$$0.06 \times 1,000,000 = 60,000 \text{ mg/L}$$

PERIMETER, CIRCUMFERENCE, AREA, AND VOLUME

Air sampling and air pollution control technologies may require you to be able to perform calculations to determine circumference, area, and volume of ventilation systems (ducting), tanks, vessels, and other structures, and perimeter of landscapes as part of field data necessary to determine the result or correct parameters to set or measure in any gaseous or volatile chemical spill mitigation action. To aid in performing these calculations, the following definitions are provided:

Area—The area of an object, measured in square units.
Base—The term used to identify the bottom leg of a triangle, measured in linear units.
Circumference—The distance around an object, measured in linear units. When determined for other than circles, it may be called the *perimeter* of the figure, object, or landscape.
Cubic units—Measurements used to express volume, cubic feet, cubic meters, etc.
Depth—The vertical distance from the bottom of the tank to the top. Normally measured in terms of liquid depth and given in terms of side wall depth (SWD), measured in linear units.
Diameter—The distance from one edge of a circle to the opposite edge passing through the center, measured in linear units.
Height—The vertical distance from the base or bottom of a unit to the top or surface.
Length—The distance from one end of an object to the other, measured in linear units.

Linear units—Measurements used to express distances: feet, inches, meters, yards, etc.

Pi, π—A number in the calculations involving circles, spheres, or cones. π = 3.1416.

Radius—The distance from the center of a circle to the edge, measured in linear units.

Sphere—A container shaped like a ball.

Square units—Measurements used to express area, square feet, square meters, acres, etc.

Volume—The capacity of the unit, how much it will hold, measured in cubic units (cubic feet, cubic meters) or in liquid volume units (gallons, liters, million gallons).

Width—The distance from one side of the tank to the other, measured in linear units.

PERIMETER

Air science practitioners may be called upon to make certain measurements in the field related to air pollution control technology, or to make design considerations and other related activities. On occasion, it is necessary to determine the distance around grounds or landscapes. To measure the distance around property, buildings, and basin-like structures, you must first determine either perimeter or circumference. The *perimeter* is how far it is around an object or area, such as piece of ground. Distance is linear measurement, which defines the length along a line. Standard units of measurement (inches, feet, yards, and miles) and metric units (centimeters, meters, and kilometers) are used.

The perimeter of a rectangle (a four-sided figure with four right angles; see Figure 3.2) is obtained by adding the lengths of the four sides:

$$\text{Perimeter} = L_1 + L_2 + L_3 + L_4 \qquad (3.8)$$

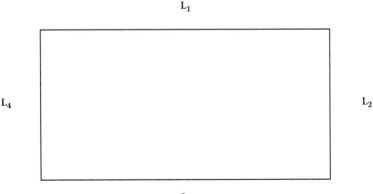

FIGURE 3.2 Perimeter.

EXAMPLE 3.11

Find the perimeter of the following rectangle:

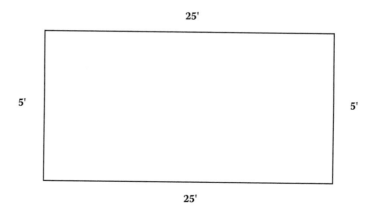

25'

5'   5'

25'

Solution

$$P = 25' + 5' + 25' + 5'$$
$$P = 60'$$

CIRCUMFERENCE

Circumference is the distance around a circle or circular object. The circumference of a circle is found by multiplying pi (π) times the *diameter* (D) (diameter is a straight line passing through the center of a circle—the distance across the circle).

$$C = \pi D \qquad (3.9)$$

where
  C = circumference
  π = Greek letter pi
  π = 3.1416
  D = diameter

EXAMPLE 3.12

A circular chemical holding tank has a diameter of 16 m. What is the circumference of this tank?

$$C = \pi(D)$$
$$C = (3.14)(\text{diameter})$$
$$C = (3.14)(16 \text{ m})$$
$$C = 50.2 \text{ m}$$

EXAMPLE 3.13

A ventilation air test inlet opening has a diameter of 3 inches. What is the circumference of the inlet opening in inches?

$$C = \pi(D)$$
$$C = 3.14 \times 3 \text{ in.}$$
$$= 9.42 \text{ in.}$$

AREA

For area measurements in air science work, three basic shapes are particularly important: circles, rectangles, and triangles.

Area is the amount of surface an object contains or the amount of material it takes to cover the surface. The area on top of a chemical tank is called the *surface area*. The area of the end of a ventilation duct is called the *cross-sectional area* (the area at right angles to the length of ducting). Area is usually expressed in square units, such as square inches (in.^{2}) or square feet (ft^2). Land may also be expressed in terms of square miles (sections) or acres (43,560 ft^2) or in the metric system as hectares.

The area of a rectangle is found by multiplying the length (L) times width (W).

$$Area = L \times W \tag{3.10}$$

EXAMPLE 3.14

Find the area of the following rectangle:

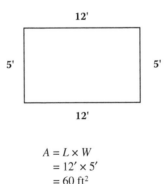

$$A = L \times W$$
$$= 12' \times 5'$$
$$= 60 \text{ ft}^2$$

The surface area of a circle is determined by multiplying π times the *radius* squared. Radius, designated r, is defined as a line from the center of a circle or sphere to the circumference of the circle or surface of the sphere.

$$A = \pi r^2 \tag{3.11}$$

where
  A = area
  π = Greek letter pi (π = 3.14)
  r = radius of a circle (radius is one-half the diameter)

EXAMPLE 3.15

What is the area of a circle with a 12-inch diameter?

$$\text{Area of circle} = \pi r^2$$
$$= \pi 6^2$$
$$= 3.14 \times 36$$
$$= 113 \text{ ft}^2$$

VOLUME

The amount of space occupied by or contained in an object, *volume*, is expressed in cubic units, such as cubic inches (in.³), cubic feet (ft³), acre feet (1 acre foot = 43,560 ft³), etc.

The volume of a rectangular object is obtained by multiplying the length times the width times the depth or height:

$$V = L \times W \times H \tag{3.12}$$

where

$L = $ length
$W = $ width
D or $H = $ depth or height

EXAMPLE 3.16

Find the volume in cubic feet of a holding pond with the following dimensions: length, 15 feet; width, 7 feet; and depth, 9 feet.

$$V = L \times W \times D$$
$$= 15' \times 7' \times 9'$$
$$= 945 \text{ ft}^3$$

For air science practitioners, representative surface areas are most often rectangles, triangles, circles, or a combination of these. Practical volume formulas used in air science calculations are given in Table 3.3.

EXAMPLE 3.17

Find the volume of a 4-inch round air duct that is 300 feet long.

Step 1: Change the diameter of the duct from inches to feet by dividing by 12:

$$D = 4 \div 12 = 0.33 \text{ ft}$$

Step 2: Find the radius by dividing the diameter by 2:

$$r = 0.33 \text{ ft} \div 2 = 0.165$$

TABLE 3.3
Volume Formulas

Sphere volume	$= (\pi/6)(\text{diameter})^3$
Cone volume	$= 1/3$ (volume of a cylinder)
Rectangular tank volume	$=$ (area of rectangle) (D or H)
	$=$ (LW)(D or H)
Cylinder volume	$=$ (area of cylinder)(D or H)
	$= \pi r^2$ (D or H)
Triangle volume	$=$ (area of triangle)(D or H)
	$=$ (bh/2)(D or H)

Step 3: Find the volume:

$$V = L \times \pi r^2$$
$$V = 300 \text{ ft} \times \pi \times (.0225) \text{ ft}^2$$
$$V = 21.2 \text{ ft}^2$$

EXAMPLE 3.18

Find the volume of a smokestack that is 36 inches in diameter (entire length) and 96 inches tall.

Step 1: Find the radius of the stack. The radius is one-half the diameter:

$$36 \text{ in.} \div 2 = 18 \text{ in.}$$

Step 2: Find the volume:

$$V = H \times \pi r^2$$
$$V = 96 \text{ in.} \times \pi(18)^2$$
$$V = 96 \text{ in.} \times \pi(324 \text{ in.}^2)$$
$$V = 97,667 \text{ in.}^3$$

PLANE GEOMETRY (EPA, 2007)

In conjunction with the basic geometric operations discussed above, it is also important to discuss and describe basic applications of plane geometry in air science. Geometry is needed in many air pollution control system components. For example, plane geometry is used to evaluate the velocities through ductwork and stacks, the minimum and maximum sizes of control systems, and the areas available for heat transfer in condensers. Solid geometry is used to evaluate the capacity of hoppers and the residence time of combustion gases in incinerators.

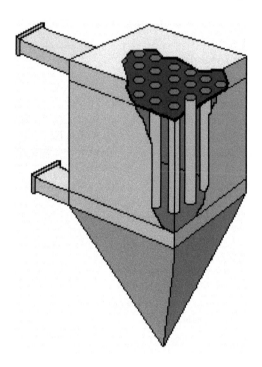

FIGURE 3.3 From EPA (2007).

To gain appreciation for and understanding of applications of plane geometry in air pollution control, consider the following question: What three-dimensional geometric shapes do you see in the fabric filter shown in Figure 3.3? For the answer, see Figure 3.4, which clearly shows the three-dimensional geometric shapes in the fabric filter.

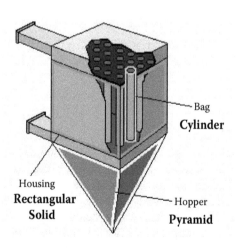

FIGURE 3.4 From EPA (2007).

MATERIAL BALANCE

Material balances are one of the most basic and useful tools in the air pollution engineering field. Stated simply, a materials balance means what goes in, must come out. Matter is neither created nor destroyed in industrial processes (nonradioactive only).

$$\text{Mass}_{(Out)} = \text{Mass}_{(In)} \qquad (3.13)$$

Material balances are used in a wide variety of air pollution control calculations. For example, they are used to evaluate the following:

- Formation of combustion products in boilers
- Rates of air infiltration into air pollution control systems
- Material requirements for process operations
- Rate of ash collection in air pollution control systems
- Humidities of exhaust gas streams
- Exhaust gas flow rates form multiple sources controlled by a single air pollution control system
- Gas flow rates from combustion processes

This principle, called the conservation of matter, can be applied in solving problems involving the quantities of matter moving in various parts of a process.

FORCE, WEIGHT, AND MASS

In air science work, the properties of gases must be known for proper selection and operation of air pollution control devices and ancillary equipment (ductwork). Along with knowing the properties of gases, the air science practitioner must also be familiar with and understand their meaning as well as the difference between the terms *force*, *weight*, and *mass*.

Force is any influence that tends to change the state of rest or the uniform motion in a straight line of a body. Stated in simpler terms, force is a push or a pull exerted on an object to change its position or movement, including starting, stopping, and changing its speed or direction of movement. In a compressed air (pneumatic) system, force must be present at all times for the system to function.

A substance or object has *weight* depending on its *mass* (represents the amount of matter in an object, and its *inertia*, or resistance to movement) and the strength of the earth's gravitational pull, which decrease with height. Consequently, an object weighs less at the top of a mountain than at sea level. An object's inertia determines how much force is needed to lift or move the object or to change its speed or direction of movement.

Another important physical property is *density*, which is a scaler quantity. The density of an object is its weight for a specific volume or unit of measure, a measure of the compactness of a substance. Density is equal to its mass per unit volume and is measured in kilogram per cubic meter or pounds per cubic foot. The density of a mass *m* occupying a volume *V* is given by the formula

$$D = m/V \qquad (3.14)$$

TABLE 3.4
Densities of Gases @ STP (Standard Temperature and Pressure: 0°C and 1 atm)

Air	1.3
Hydrogen	0.09
Helium	0.18
Methane	0.72
Nitrogen	1.25
Oxygen	1.43
Carbon dioxide	1.98
Propane	2.02
Butane	2.65

The density of a cubic foot of dry air at atmospheric pressure and a temperature of 60°F is 0.076 pounds and is more commonly expressed as 0.076 lb/ft^3. The density of wet air at atmospheric pressure with 100% relative humidity and a temperature of 60°F is 0.075 lb/ft^3. Humid air is less dense than dry air because the water vapor will not allow the air to compress as much. As a result, humid air weighs less. Air's relatively low density makes it suitable for long-distance and high-speed control applications in pneumatic systems.

The densities of some common gases are given in Table 3.4.

PRESSURE

Pressure is the amount of force (in pounds) exerted on an object or a substance, divided by the area (in square inches) over which this force is exerted. Pressure can be measured and specified in different ways but is commonly measured in pounds per square inch (psi). The SI unit of pressure is the pascal (newton per square meter), equal to 0.01 millibars. At the edge of earth's atmosphere, pressure is zero, whereas at sea level atmospheric pressure due to weight of the air above is about 100 kilopascals (1,013 millibars or 1 atmosphere).

GAS PRESSURE

The pressures of the gas streams throughout the particulate control system are very important because gas pressure data are often used to evaluate operating conditions. The total pressure of a gas stream is the sum of the static pressure and velocity pressure of the gas stream.

$$TP = SP + VP \qquad (3.15)$$

where
    TP = total pressure
    SP = static pressure
    VP = velocity pressure

Velocity pressure is exerted only in gas streams that are in motion. This part of the total pressure is of concern only during emission tests and gas flow rate measurements and is not routinely monitored by plant personnel.

Static pressure is the pressure exerted by all gases. This pressure is basically related to the number of gas molecules in a given volume and at a given temperature. If the number of molecules in the space increases, the pressure increases. An increase in the gas temperature increases the kinetic energy of the molecules, and the static pressure increases (EPA, 2007).

PRESSURE SCALES

Like temperature, gas pressure can be expressed in both relative and absolute terms. The absolute pressure scale starts at zero gas pressure (no molecules—a vacuum) and has no practical maximum limit. The absolute temperature and pressure scales are most useful for the scientific and engineering calculations necessary to evaluate the following:

- Hoods and ventilation systems
- Source emission rates
- Air pollution control equipment performance

Absolute pressure is used whenever it is necessary to use the ideal gas laws to calculate gas flow rates. However, in the air pollution control and emission measurement fields, gas pressures are often monitored in terms of relative pressures, and it is necessary to convert the data to absolute pressure prior to performing the calculations.

ATMOSPHERIC PRESSURE

In this text we are primarily concerned with *atmospheric pressure* (i.e., the static pressure exerted by ambient air), but it should be noted that two other kinds of pressure—below atmospheric and pneumatic system pressure—are also common. Atmospheric pressure is an absolute pressure because it is directly related to the number of molecules and their kinetic energy. Atmospheric pressure at sea level equals 14.7 psi; pressure is lower above sea level and higher below sea level.

The complete or partial absence of air (indicating below atmospheric pressure) is often referred to as a *vacuum* or partial vacuum. In some applications it may also be called a negative or suction pressure. Vacuum is normally measured using special gauges or with a column of mercury. When all the air above the column is evacuated, atmospheric pressure is exerted on the pool of mercury below the tube. This pressure raises the column to a height of approximately 30 inches. In most applications a vacuum is measured in inches of mercury instead of psi.

Note that most pressure gauges in a pressurized air system measure only pressure that is higher than the atmospheric pressure surrounding them. You may have noticed that when a pressure gauge is disconnected, it reads zero pounds per square inch, which is known as *gauge pressure* (0 psig). For example, a reading of 300 on an air system pressure gauge tells you that the air pressure is 300 psi above atmospheric. If we add atmospheric pressure to this gauge pressure, the total pressure is 314.7 pounds per square inch (300 + 14.7), which is known as *absolute pressure* (psia). Remember that although absolute pressure readings are important in some pressurized air system calculations, the distinction between psig and psia is usually unimportant in the average air system. As a result, gauge pressure readings are usually expressed in psi.

WORK AND ENERGY

Work is the transference of energy that occurs when a force is applied to a body that is moving in such a way that the force has a component in the direction of the body's motion. Stated in simpler fashion: work takes place when a force (in pounds or newtons) moves through a distance (in inches, feet, or meters). The amount of work done is expressed in the English system of measurement in foot-pounds or inch-pounds, as shown in the following equation:

$$\text{Work} = \text{Force (pounds)} \times \text{Distance (feet or inches)} \qquad (3.16)$$
$$= \text{Foot-pounds or inch-pounds}$$

Pascal's law states that *when there is an increase in pressure at any point in a confined fluid (air), there is an equal increase at every other point in the container.* A container, as shown in Figure 3.5, contains a fluid. There is an increase in pressure as the length of the column of liquid increases, due to the increased mass of the fluid above. For example, in Figure 3.5, P3 would be the highest value of the three pressure readings, because it has the highest level of fluid above it. If the container in Figure 3.5 had an increase in overall pressure, that same added pressure would affect each of the gauges (and the liquid throughout) the same. For example P1, P2, and P3 were originally 1, 3, and 5 units of pressure, and 5 units of pressure were added to the system; the new readings would be 6, 8, and 10.

Applied to a more complex system shown in Figure 3.6, Pascal's law allows forces to be multiplied. The cylinder on the left shows a cross section area of 1 square inch, while the cylinder on the right shows a cross section area of 10 square inches. The cylinder on the left has a weight (force) of 1 pound acting downward on the

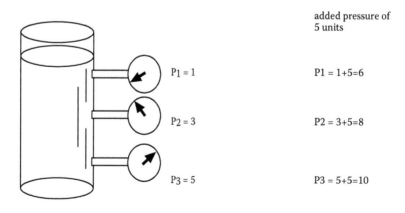

FIGURE 3.5 From EPA (2007).

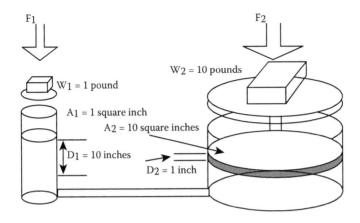

FIGURE 3.6 From EPA (2007).

piston, which lowers the fluid 10 inches. As a result of this force, the piston on the right lifts a 10-pound weight a distance of 1 inch.

The 1-pound load on the 1-square-inch area causes an increase in pressure on the fluid in the system. This pressure is distributed equally throughout and acts on every square inch of the 10-square-inch area of the large piston. As a result, the larger piston lifts up a 10-pound weight. The larger the cross section area of the second piston, the larger the mechanical advantage, and the more weight it lifts (NASA, 2007).

Power is defined as the time rate of doing work—or the amount of work (foot-pounds) done in a given length of time (seconds or minutes), or foot-pounds per minute. The following equation is used to determine the amount of power:

$$\text{Power (P)} = \frac{\text{Work}}{\text{Time}}$$

$$= \frac{\text{Foot-pounds}}{\text{Seconds (or minutes)}}$$

To convert foot-pounds to inch-pounds you should multiply by 12.

Note that for the amount of power calculated to be meaningful, it must be compared with a unit of measurement. The common unit of power measurement is *horsepower*, calculated as follows:

$$1 \text{ hp} = \frac{33,000 \text{ ft-lb}}{\text{Seconds (or minutes)}} \tag{3.17}$$

When power is used to perform work, energy is expended. The law of conservation of energy states that *energy cannot be created or destroyed. It can only be transformed.* Thus, we use one kind of energy to get other kinds of energy. Some of this energy does useful work while some of it is wasted (as heat energy—remember, energy cannot be destroyed or lost) in overcoming friction.

DIFFUSION AND DISPERSION

Diffusion can be described as the spontaneous and random movement of molecules or particles in a gas (or liquid) from a region in which they are at a high concentration, until a uniform concentration is achieved throughout. No mechanical mixing or stirring is involved. This should not be confused with evaporation, which is the changing of a liquid to a gas.

   Dispersion can be described as the temporary mixing of liquid particles with a gas.

   Diffusion and dispersion are important in air pollution. For example, in dispersion, air pollutants are diluted and reduced in concentration. Air pollution dispersion mechanisms are a function of the prevailing meteorological conditions. Diffusion and dispersion (in air pollution) will be discussed much more fully later.

COMPRESSIBILITY

Air, unlike liquids, is readily compressible, and large quantities can be stored in relatively small containers. The more the air is compressed, the higher its pressure becomes. The higher the pressure in a container, the stronger the container must be. Gases are important compressible fluids, not only from the standpoint that a gas can be a pollutant, but also because gases convey the particles (particulate matter) and gaseous pollutants (Hesketh, 1991).

GAS LAWS

Gases can be pollutants as well as the conveyors of pollutants. Air (which is mainly nitrogen) is usually the main gas stream. Gas conditions are usually described in two ways: *standard temperature and pressure* (STP) and *standard conditions* (SC). STP represents 0°C (32°F) and 1 atm. SC is more commonly used and represents typical room conditions of 20°C (70°F) and 1 atm; SC is usually measured in cubic meters, Nm^3, or standard cubic feet (scf).

   To understand the physics of air it is imperative to have an understanding of various physical laws that govern the behavior of pressurized gases. One of the more well-known physical laws, mentioned earlier, is *Pascal's law*. In addition to one of its previously mentioned parameters, Pascal's law also states that *a confined gas (fluid) transmits externally applied pressure uniformly in all directions, without change in magnitude.* This parameter can be seen in a container that is flexible; it will assume a spherical (balloon) shape. However, you probably have noticed that most compressed-gas tanks are cylindrical in shape (which allows use of thinner sheets of steel without sacrificing safety) with spherical ends to contain the pressure more effectively.

BOYLE'S LAW

Though gases are compressible, note that for a given mass flow rate, the actual volume of gas passing through the system is not constant within the system due to changes in pressure. This physical property (the basic relationship between the pressure of a gas and its volume) is described by Boyle's law (named for its discoverer: Irish physicist and chemist Robert Boyle in 1662), which states that *the absolute pressure of a confined quantity of gas varies inversely with its volume, if its temperature does not change*. For example, if the pressure of a gas doubles, its volume will be reduced by half, and vice versa. That is, *as pressure goes up, volume goes down*, and vice versa. This means, for example, that if 12 ft³ of air at 14.7 psia is compressed to 1 ft³, air pressure will rise to 176.4 psia, as long as air temperature remains the same. This relationship can be calculated as follows:

$$P_1 \times V_1 = P_2 \times V_2 \qquad\qquad (3.18)$$

where
   P_1 = original pressure (units for pressure must be absolute)
   P_2 = new pressure (units for pressure must be absolute)
   V_1 = original gas volume at pressure P_1
   V_2 = new gas volume at pressure P_2

This equation can also be written as

$$\frac{P_2}{P_1} = \frac{V_1}{V_2} \text{ or } \frac{P_1}{P_2} = \frac{V_2}{V_1}$$

To allow for the effects of atmospheric pressure, always remember to convert from gauge pressure *before* solving the problem, then convert back to gauge pressure *after* solving it:

$$\text{psia} = \text{psig} + 14.7 \text{ psi}$$

and

$$\text{psig} = \text{psia} - 14.7 \text{ psi}$$

Note that in a pressurized gas system where gas is caused to move through the system by the fact that gases will flow from an area of high pressure to one of low pressure, we will always have a greater actual volume of gas at the end of the system than at the beginning (assuming the temperature remains constant).

   Let us take a look at a typical gas problem using Boyle's law:

EXAMPLE 3.19
What is the gauge pressure of 12 ft³ of air at 25 psig when compressed to 8 ft³?

Solution

$$25 \text{ psig} + 14.7 \text{ psi} = 39.7 \text{ psia}$$
$$P_2 = P_1 \times \frac{V_1}{V_2} = 39.7 \times \frac{12}{8} = 59.6 \text{ psia}$$

$$\text{psig} = \text{psia} - 14.7 \text{ psi}$$
$$= (59.6 \text{ psia}) - (14.7 \text{ psi}) = 44.9 \text{ psig}$$

The gauge pressure is 44.9 psig (remember that the pressures should always be calculated on the basis of absolute pressures instead of gauge pressures).

CHARLES'S LAW

Another physical law dealing with temperature is Charles's law (discovered by French physicist Jacques Charles in 1787). It states that *the volume of a given mass of gas at constant pressure is directly proportional to its absolute temperature* (the temperature in Kelvin (273 + °C) or Rankine (absolute zero = –460°F, or 0°R).)

This is calculated by using the following equation:

$$P_2 = P_1 \times \frac{T_2}{T_1}$$
(3.19)

Charles's law also states *if the pressure of a confined quantity of gas remains the same, the change in the volume (V) of the gas varies directly with a change in the temperature of the gas*, as given in the following equation:

$$V_2 = V_1 \times \frac{T_2}{T_1}$$
(3.20)

IDEAL GAS LAW

The ideal gas law combines Boyle's and Charles's laws because air cannot be compressed without its temperature changing. The ideal gas law is expressed by the following equation:

$$\frac{P_1 \times V_1}{T_1} = \frac{P_2 \times V_2}{T_2}$$
(3.21)

Note that the ideal gas law is still used as a design equation even though the equation shows that the pressure, volume, and temperature of the second state of a gas are equal to the pressure, volume, and temperature of the first state. In actual practice, however, other factors, such as humidity, heat of friction, and efficiency losses, all affect the gas. Also, this equation uses absolute pressure (psia) and absolute temperatures (°R) in its calculations.

In air science practice, the importance of the ideal gas law cannot be overstated. It is one of the fundamental principles used in calculations involving gas flow in air pollution–related work. This law is used to calculate actual gas flow rates based on the quantity of gas present at standard pressures and temperatures. It is also used to determine the total quantity of that contaminant in a gas that can participate in a chemical reaction. Key parameters used in ideal gas law are:

- Number of moles of gas
- Absolute temperature
- Absolute pressure

In practical applications, practitioners generally use the following standard ideal gas law equation:

$$V = \frac{nRT}{P} \text{ or } PV = nRT \qquad (3.22)$$

where

V = volume
n = number of moles
R = universal gas constant
T = absolute temperature
P = absolute pressure

Did You Know?

Moles is a measure of the number of molecules present. The value of R depends on the units used for the other parameters.

EXAMPLE 3.20

What is the volume of 1 pound mole (denoted lb mole) of combustion gas at an absolute pressure of 14.7 psia and a temperature of 68°F? (These are EPA-defined standard conditions.)

Solution

$$V = \frac{nRT}{P}$$

1. Convert the temperature from relative to absolute scale (from °F to °R).

$$T_{Absolute} = 68°F + 460 = 528°R$$

2. Calculate the gas volume.

$$V = \frac{1\text{lb mole} \times \dfrac{10.73\,(\text{psia})\,(\text{ft}^3)}{(\text{lb mole})(°R)} \times 528°R}{14.7\text{ psia}}$$

$$= 385.4 \text{ ft}^3$$

FLOW RATE

Gas flow rate is a measure of the volume of gas that passes a point in an industrial system during a given period of time. The ideal gas law tells us that this gas flow rate varies depending on the temperature and pressure of the gas stream and the number of moles of gas moving per unit of time.

When gas flow rates are expressed at actual conditions of temperature and pressure, the actual gas flow rate is being used. As you will learn later, gas flow rates can also be expressed at standard conditions of temperature and pressure; this is referred to as the standard gas flow rate.

GAS CONVERSIONS

Gases of interest in air pollution control are usually mixtures of several different compounds. For example, air is composed of three major constituents: nitrogen (N_2) at approximately 78.1%, oxygen (O_2) at approximately 20.9%, and argon at 0.9%. Many flue gas streams generated by industrial processes consist of the following major constituents: (1) nitrogen, (2) oxygen, (3) argon, (4) carbon dioxide (CO_2), and (5) water vapor (H_2O). Both air and industrial gas streams also contain minor constituents, including air pollutants, present at concentrations that are relatively low compared to these major constituents.

Did You Know?

Argon is usually not listed in most industrial gas analyses because it is chemically inert and difficult to measure. The argon concentration is often combined with the nitrogen concentration to yield a value of 79.0%.

There is a need for ways to express both the concentrations of the major constituents of the gas stream and the concentrations of the pollutants present as minor constituents at relatively low concentrations. There are a variety of ways to express gas phase concentrations, which can easily be converted from one type of units to another. They include the following.

MAJOR CONSTITUENTS

- *Volume percent*—One of the most common formats used to express the concentrations of major gas stream constituents such as oxygen, nitrogen, carbon dioxide, and water vapor. The format is very common partially because the gas stream analysis techniques used in EPA emission testing methods provide data directly in a volume percent format.
- *Partial pressure*—Concentrations can also be expressed in terms of partial pressures. This expression refers to the part of the total pressure exerted by one of the constituent gases.

Gases composed of different chemical compounds such as molecular nitrogen and oxygen behave physically the same as gases composed of a single compound. At any give temperature, one mole of a gas exerts the same pressure as one mole of any other type of gas. All of the molecules move at a rate that is dependent on the absolute temperature, and they exert pressure. The total pressure is the sum of the pressures of each of the components. The equations below are often called Dalton's law of partial pressures.

$$P_{Total} = p_i + p_{ii} + p_{iii} \cdots p_n \tag{3.23}$$

$$P_{Total} = \sum_{i=1}^{n} p_i \tag{3.24}$$

$$\text{Partial pressure (gas)} = \left[\frac{\text{Volume \% (gas)}}{100 \%} \right] \times P_{Total} \tag{3.25}$$

Because the partial pressure value is related to the total pressure, concentration data expressed as partial pressure are not the same at actual and standard conditions. The partial pressure values are also different; they are in American engineering units and CGS (centimeter-gram-second) units.

BOTH MAJOR AND MINOR CONSTITUENTS

- *Mole fraction*—An expression of the number of moles of a compound divided by the total number of moles of all the compounds present in the gas.

MINOR CONSTITUENTS

- Parts per million (ppm)
- Milligrams per cubic meter (mg/m^3)
- Micrograms per cubic meter ($\mu g/m^3$)
- Nanograms per cubic meter (ng/m^3)

All of the concentration units above can be expressed in a dry format as well as corrected to a standard oxygen concentration. These corrections are necessary because moisture and oxygen concentrations can vary greatly in gas streams, causing variations in pollutant concentrations.

GAS VELOCITY

Gas velocity is one of the fundamental design variables for ventilation systems and air pollution control equipment. Gas streams containing particulate are usually maintained at velocities of 3,000 to 4,500 ft/min in ductwork leading to particulate collectors to minimize particle deposition. The velocity of gas streams without particulate matter is often in the range of 1,500 to 3,000 ft/min. The gas velocities in air pollution control equipment are usually low to allow for sufficient time to remove the contaminants. For example, gas velocities through electrostatic precipitators are usually in the range of 2.5 to 6 ft/s. The filtration velocities through pulse jet fabric filters are usually in the range of 2 to 10 ft/min. Variations in the gas velocity can have a direct impact on the contaminant removal efficiency.

The average velocity of a gas stream in an emission testing probe, an industrial duct, or an air pollution control device is a function of the actual gas flow rate and the cross-sectional flow area.

$$v = \text{Gas velocity} = \frac{\text{Gas flow rate, actual}}{\text{Area}} \qquad (3.26)$$

GAS STREAM TREATMENT (RESIDENCE) TIME

The flow rate of the gas stream through an air pollution control system determines the length of time that the pollutants can be removed from the gas stream. This is termed the *treatment time* or *residence time*. These common equipment sizing parameters are defined mathematically in Equation 3.27.

$$\text{Treatment time} = \text{Residence time}$$
$$= \frac{\text{Volume of control device}}{\text{Gas flow rate, actual}} \qquad (3.27)$$

GAS DENSITY

Gas density is important primarily because it affects the flow characteristics of the moving gas streams. Gas density affects the velocities of gas through ductwork and air pollution control equipment. It determines the ability to move the gas stream using a fan. Gas density affects the velocities of gases emitted from the stack, and thereby influences the dispersion of the pollutants remaining in the stack gases. It affects the ability of particles to move through gases. It also affects emission testing. Gas density data are needed in many of the calculations involved in air pollution control equipment evaluation, emission testing, and other air pollution control–related studies.

As discussed earlier, the volume of a gas increases as the temperature increases due to the motion of the gas molecules. As the volume occupied by the gas increases, its density decreases. Density is the mass per unit volume, as indicated in Equation 3.28.

$$P_{(T=i,\ P=j)} = \frac{m}{V_{(T=I,\ P=j)}} \qquad (3.28)$$

where
$P_{(T=I,\ P=j)}$ = density at $T = I$, $P = j$
M = mass of a substance
$V_{(T=I,\ P=j)}$ = volume at $T = I$, $P = j$
T = absolute temperature
P = absolute pressure

Did You Know?

Gas density is expressed as the mass per unit of volume of gas. The gas volume is always expressed at actual conditions. The gas volume is not corrected for temperature, pressure, moisture, or oxygen levels.

HEAT CAPACITY AND ENTHALPY

The heat capacity of a gas is the amount of heat required to change the temperature of a unit mass of gas one temperature degree.

Enthalpy represents the total quantity of internal energy, such as heat, measured for a specific quantity of material at a given temperature. Enthalpy data are often represented in units of energy (e.g., Btu, kcal, joule, etc.). The enthalpy content change is often expressed in Btu/unit mass (Btu/lb$_m$) or Btu/unit time (Btu/scf). The change in enthalpy of the total quantity of material present in a system is expressed in units of Btu/unit time (Btu/min). The symbols H and ΔH denote enthalpy and the change in enthalpy, respectively.

HEAT AND ENERGY IN THE ATMOSPHERE

In addition to the importance of heat on a particular air stream, heat also has an impact on earth's atmosphere, and thus on atmospheric science. The sun's energy is the prime source of the earth's climatic system. From the sun, energy is reflected, scattered, absorbed, and reradiated within the system but without uniform distribution. Some areas receive more energy than they lose; in some areas the reverse occurs. If this situation were able to continue for long, the areas with an energy surplus would get hotter—too hot, and those with a deficit would get colder—too cold. This does not happen because the temperature differences produced help to drive the wind and ocean currents of the world. They carry heat with them, either in the sensible or latent forms, and help to counteract the radiation imbalance. Winds from the tropics are therefore normally warm, carrying excess heat with them. Polar winds are blowing from areas with a deficit of heat and so are cold. Acting together, these energy transfer mechanisms help to produce the present climates on earth.

ADIABATIC LAPSE RATE

The atmosphere is restless, always in motion either horizontally or vertically, or both. As air rises, pressure on it decreases and in response it expands. The act of expansion to encompass its new and larger dimensions requires an expenditure of energy; since temperature is a measure of internal energy, this use of energy makes its temperature drop—this is an important point—an important process in physics (especially in air physics).

This phenomenon is known as the *adiabatic lapse rate*. Simply, *adiabatic* refers to a process that occurs with or without loss of heat, especially the expansion or contraction of a gas in which a change takes place in the pressure or volume, although no heat is allowed to enter or leave.

Lapse rate refers to the rate at which air temperature decreases with height. The normal lapse rate in stationary air is on the order of 3.5°F/1,000 ft (6.5°C/km). This value may vary with latitude and changing atmospheric conditions (e.g., seasonal changes). A parcel of air that is not immediately next to the earth's surface is sufficiently well insulated by its surroundings that either expansion or compression of the parcel may be assumed to be adiabatic.

The air temperature may be calculated for any height by the general formula

$$T = T_0 - Rh \tag{3.29}$$

where

T = temperature of the air
h = height of air
T_0 = temperature of the air at the level from which the height is measured
R = lapse rate

EXAMPLE 3.21

If the air temperature of stationary air (R = 3.5°F/1,000 ft) at the earth's surface is 70°F, then at 5,000 feet the stationary air temperature would be

$$
\begin{aligned}
T &= T_0 - Rh \\
&= 70°F - (3.5°F/1{,}000 \text{ ft})(5{,}000 \text{ ft}) \\
&= 70°F - 17.5°F = 52.5°F
\end{aligned}
$$

The formula simply says that for every 1,000 feet of altitude (height), 3.5° is subtracted from the initial air temperature, in this case.

Adiabatic lapse rates have an important relationship with atmospheric stability and will be discussed in greater detail later in the text.

VISCOSITY

All fluids (gases included) resist flow. *Absolute viscosity* is a measure of this resistance to flow. The absolute viscosity of a gas for given conditions may be calculated from the following formula:

$$\mu = 51.12 + 0.372(T) + 1.05 \times 10^{-4}(T)^2 \tag{3.30}$$
$$+ 53.147 \, (\% \, O_2/100\%) - 74.143 \, (\% \, H_2O/100\%)$$

where

μ = absolute viscosity of gas at the prevailing conditions, micropoise
T = gas absolute temperature, K
$\% \, O_2$ = oxygen concentration, % by volume
$\% \, H_2O$ = water vapor concentration, % by volume

As this equation indicates, the viscosity of a gas increases as the temperature increases. It is harder to push something (e.g., particles) through a hot gas stream than a cooler one due to increased molecular activity as temperature rises, which results in increased momentum transfer between the molecules. For liquids, the opposite relationship between viscosity and temperature holds. The viscosity of a liquid decreases as temperature increases. It is harder to push something through a cold liquid than a hot one because in liquids, hydrogen bonding increases with colder temperatures.

Did You Know?

Gas viscosity actually increases very slightly with pressure, but this variation is very small in most air pollution–related engineering calculations.

Gas viscosity actually increases very slightly with pressure, but this variation is very small in most air pollution–related engineering calculations.

The absolute viscosity and density of a gas are occasionally combined into a single parameter since both of these parameters are found in many common equations describing gas flow characteristics. The combined parameter is termed the kinematic viscosity. It is defined in Equation 3.31.

$$u = \mu/p \tag{3.31}$$

where
u = kinematic viscosity, m^2/s
μ = absolute viscosity, Pa·s
p = gas density, gm/cm^3

The kinematic viscosity can be used in equations describing particle motion through gas streams. The expression for kinematic viscosity is used to simplify these calculations.

FLOW CHARACTERISTICS

When fluids such as gases are moving slowly, the bulk material moves as distinct layers in parallel paths, as illustrated in Figure 3.7. The only movement across these layers is the molecular motion, which creates viscosity. This is termed *laminar flow*.

As the velocity of the gas stream increases, the bulk movement of the gas changes. Eddy currents develop that cause mixing across the flow stream. This is called *turbulent flow* and is essentially the only flow characteristic that occurs in air pollution control equipment and emission testing–related situations. Turbulent flow is illustrated in Figure 3.8.

A dimensionless parameter called the *Reynolds number* is used to characterize fluid flow. It is the ratio of the inertial force that is causing gas movement to the viscous force that is restricting movement. The Reynolds number is calculated using Equation 3.32. Consistent units must be used to ensure that the Reynolds number is dimensionless.

$$N_{Re(g)} = Lvp/\mu \tag{3.32}$$

FIGURE 3.7 Laminar (streamline) flow.

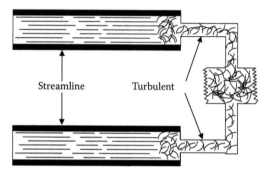

FIGURE 3.8 Turbulent flow.

where

  L = linear dimensions (usually duct diameter)
  v = fluid velocity
  p = fluid density
  μ = fluid viscosity

Reynolds numbers less than 2,000 are associated with laminar flow conditions. Due to the relatively low velocities associated with this type of flow, they are rarely encountered in air pollution field situations.

Reynolds numbers above 10,000 are associated with turbulent flow. In many field situations, the Reynolds numbers exceed 100,000.

Did You Know?

Between Reynolds numbers of 2,000 and 10,000, turbulent flow conditions have not fully developed. This is called the transitional flow range. Flow in this transitional range has characteristics of both laminar and turbulent flow.

Essentially, all gas flow situations are turbulent in air pollution control systems, emission test and monitoring equipment, and dispersion modeling studies. However, this does not mean that the gas stream is entirely well mixed. In reality, the side-to-side mixing (and even mixing in the direction of flow) can be limited. For this reason, it is possible to have different concentrations of pollutants at different positions in the duct. This is called *pollutant stratification*. It can be caused by a variety of factors: combining of two separate gas streams into a single duct, temperature differences in the gas stream, and in-leakage of cold ambient air into the duct.

Stratification does not exist in most industrial gas handling systems. However, it is important to check for this condition prior to installation of continuous emission monitors or other instruments, which are located at a single sampling or measurement point in the gas stream. These measurements can be in error if stratification is severe.

THE ALBINO RAINBOW

Have you ever looked up into the sky and seen eleven suns? Have you been at sea and witnessed the towering, spectacular Fata Morgana? (Do you even know what Fata Morgana is? If not, hold on, we will get to it shortly.) How about a glory—have you ever seen one? Or how about the albino rainbow—have you seen one lately? Do you know what these are? They will be described below.

Normally, when we look up into the sky, we see what we expect to see: an ever-changing backdrop of color, with dynamic vistas of blue sky, white, puffy clouds, gray storms, and gold and red sunsets. On some occasions, however, when atmospheric conditions are just right, we can look up at the sky or out upon the horizon and see the strange phenomena (or lights in the sky) mentioned above. What causes these momentary wonders?

Because earth's atmosphere is composed of gases (air)—it is actually a sea of molecules. These molecules of air scatter the blue, indigo, and violet shorter wavelengths of light more than the longer orange and red wavelengths, which is why the sky appears blue.

What are wavelengths of light? Simply put, a wavelength of light actually refers to the electromagnetic spectrum. The portion of the spectrum visible to the human eye falls between the infrared and ultraviolet wavelengths. The colors that make up the visible portion of the electromagnetic spectrum are commonly abbreviated by the acronym ROY G BIV (red, orange, yellow, green, blue, indigo, and violet).

The word *light* is commonly given to visible electromagnetic radiation. However, only the frequency (or wavelength) distinguishes visible electromagnetic radiation from the other portions of the spectrum.

Let us get back to the sky—to looking up into the sky. Have you ever noticed that right after a rain shower, the sky appears a dark shade of blue? Have you looked out upon the horizon at night or in the morning and noticed that the sun's light gives off a red sky? This phenomenon is caused by sunlight passing through large dust particles, which scatter the longer wavelengths. Have you ever noticed that fog and cloud droplets, with diameters larger than the wavelength of light, scatter all colors equally and make the sky look white? Maybe you have noticed that fleeting greenish light that appears just as the sun sets? It occurs because different wavelengths of light are *refracted* (bent) in the atmosphere by differing amounts. Because green light is refracted more than red light by the atmosphere, green is the last to disappear.

What causes rainbows? A rainbow is really nothing more than an airborne prism. When sunlight enters a raindrop, refraction and reflection take place, splitting white light into the spectrum of colors from red to blue and making a rainbow.

Earlier a glory was mentioned. Interactions of light waves can produce a glory. For example, if you were standing on a mountain, with the sun to your back, you may cast a shadow on the fog in the valley. Your shadow may appear to be surrounded by colored halos—a glory. The glory is caused by light entering the edges of tiny droplets and being returned in the same direction from which it arrived. These light waves interfere with each other, sometimes canceling out and sometimes adding to each other.

Why do we sometimes see multiple suns? Reflection and refraction of light by ice crystals can create bright halos in the form of arcs, rings, spots, and pillars. Mock suns (sun dogs) may appear as bright spots 22° to the left or right of the sun. Sun pillars occur when ice crystals act as mirrors, creating a bright column of light extending above the sun. Such a pillar of bright light may be visible even when the sun has set.

What is the Fata Morgana? It is an illusion, a mirage, which often fools sailors into seeing mountain ranges floating over the surface of the ocean. Henry Wadsworth Longfellow had his own take on Fata Morgana, in his poem of the same title, which basically explains the essence of the phenomenon:

O sweet illusions of song
That tempt me everywhere,
In the lonely fields, and the throng
Of the crowded thoroughfare!

I approach and ye vanish away,
I grasp you, and ye are gone;
But over by night and by day,
The melody soundeth on.

As the weary traveler sees
In desert or prairie vast,
Blue lakes, overhung with trees
That a pleasant shadow cast;

Fair towns with turrets high,
And shining roofs of gold,
That vanish as he draws nigh,
Like mists together rolled —

So I wander and wander along,
And forever before me gleams
The shining city of song,
In the beautiful land of dreams.

But when I would enter the gate
Of that golden atmosphere,
It is gone, and I wonder and wait
For the vision to reappear.

An albino rainbow is an eerie phenomenon that can only be seen on rare occasions in foggy conditions.

Atmospheric phenomena (lights in the sky) are real, apparent, and sometimes visible. Awe inspiring as they are, their significance—their actual existence—is based on physical conditions that occur in our atmosphere.

REFERENCES AND RECOMMENDED READING

EPA. 2007. Basic concepts in environmental sciences: Modules 1 and 2. www.epa.gov/eogapti1/module 1 and 2 (accessed December 30, 2007).

Heinlein, R. A. 1973. *Time enough for love*. New York: G. P. Putnum's Sons.

Hesketh, H. E. 1991. *Air pollution control: Traditional and hazardous pollutants*. Lancaster, PA: Technomic Publishing Company.

NASA. 2007. Pascal's principle and hydraulics. www.grc.nasa.gov/WWW (accessed December 29, 2007).

Price, J., 1991. *Basic Math Concepts for Water and Wastewater Plant Operators*. Lancaster, PA: Technomic Publishing Company.

Spellman, F. R., and Whiting, N. 2006. *Environmental science and technology: Concepts and applications*. Rockville, MD: Government Institute.

4 Particle Physics

In air pollution control technology, it is important to have a basic understanding of not only the physics of gases but also the physics of airborne particles or particulate matter (or aerosol) that might be entrained within the gaseous stream.

INTRODUCTION

Particulates constitute a major class of air pollution. They have a variety of shapes; they include dusts, fumes, mists, and smoke; and they have a wide range of physical and chemical properties. Important particulate matter characteristics include size, size distribution, shape, density, stickiness, corrosivity, reactivity, and toxicity. With a wide range of characteristics it should be obvious that the type of collection device to be used might be better suited than others for a particular particle. For example, particulate matter typically ranges in size from 0.005 to 100 microns in diameter. This wide range in size distribution must certainly be taken into account.

CHARACTERISTICS OF PARTICLES

Understanding the characteristics of particles is important in air pollution control technology for the following reasons: The efficiency of the particle collection mechanisms strongly depends on particle size. The particle size distribution of flue gas dictates the manner in which air testing is performed and gas determines the operating conditions necessary to collect the particles. Particle characteristics are important in determining the behavior of particles in the respiratory tract.

SURFACE AREA AND VOLUME

The size range of particles of concern in air pollution studies is remarkably broad. Some of the droplets collected in the mist eliminators of wet scrubbers and the solid particles collected in large-diameter cyclones are as large as raindrops. Some of the small particles created in high-temperature incinerators and metallurgical processes can consist of less than fifty molecules clustered together. It is important to recognize the categories since particles behave differently depending on their size. Most particles of primary interest in air pollution emission testing and control have diameters that range from 0.1 to 1000 micrometers (microns).

Useful conversions for particle sizes:

$$\text{Micrometer} = \left(\frac{1}{1,000,000} \right) \text{meter} \tag{4.1}$$

TABLE 4.1
Spherical Particle Diameter, Volume, and Area

Particle Diameter (μm)	Particle Volume (cm³)	Particle Area (cm²)
0.1	5.23×10^{-16}	3.14×10^{-10}
1.0	5.23×10^{-13}	3.14×10^{-8}
10.0	5.23×10^{-10}	3.14×10^{-6}
100.0	5.23×10^{-7}	3.14×10^{-4}
1,000.0	5.23×10^{-4}	3.14×10^{2}

Source: EPA 2007.

$$\text{Micrometer} = \left(\frac{1}{10,000} \right) \text{centimeter} \qquad (4.2)$$

$$1,000 \ \mu m = 1.0 \ mm = 0.1 \ cm \qquad (4.3)$$

To appreciate the difference in particle sizes commonly found in emission testing and air pollution control, compare the diameters, volumes, and surfaces areas of particles in Table 4.1.

The data in Table 4.1 indicate that particles of 1,000 micrometers are 1,000,000,000,000 (1 trillion) times larger in volume than 0.1-micrometer particles.

As an analogy, assume that the 1,000-micrometer particle is a large domed sports stadium. A basketball in this stadium would be equivalent to a 5-micrometer particle. Approximately 100,000 spherical particles of 0.1 micrometer diameter would fit into this 5-micrometer basketball. The entire 1,000-micrometer stadium is the size of a small raindrop. As previously stated, particles over this entire size range of 0.1 to 1,000 micrometers are of interest in air pollution control (EPA, 2007).

Equations for calculating the volume and surface area of spheres are provided below.

$$\text{Surface area of a sphere} = 4\pi r^2 = \pi D^2 \qquad (4.4)$$

$$\text{Volume of a sphere} = 4\pi r^3/3 = \pi D^3/6 \qquad (4.5)$$

where
r = radius of sphere
D = diameter of sphere

EXAMPLE 4.1
Calculate the volumes of three spherical particles (in cm³) given that their actual diameters are 0.6, 6.0, and 60 micrometers.

Solution

$$V = 4/3\pi r^3$$
$$= 4.19 r^3$$

For a 0.6-micrometer particle, $r = 0.3$ micrometers $= 0.00003$ cm.

$$V = 4.19(0.00003 \text{ cm})^3$$
$$= 1.13 \times 10^{-13} \text{ cm}^3$$

For a 6.0-micrometer particle, $4 = 3.0$ micrometers $= 0.0003$ cm.

$$V = 4.19(0.0003 \text{ cm})^3$$
$$= 1.13 \times 10^{-10} \text{ cm}^3$$

For a 60-micrometer particle, $r = 30$ micrometers $= 0.003$ cm.

$$V = 4.19(0.003 \text{ cm})^3$$
$$= 1.13 \times 10^{-7} \text{ cm}^3$$

AERODYNAMIC DIAMETER

Particles emitted from air pollution sources and formed by natural processes have a number of different shapes and densities. Defining particle size for spherical particles is easy; it is simply the diameter of the particle. For nonspherical particles, the term *diameter* does not appear to be strictly applicable.

In air pollution control, particle size is based on particle behavior in the earth's gravitational field. The *aerodynamic equivalent diameter* refers to a spherical particle of unit density that falls at a standard velocity. Particle size is important because it determines atmospheric lifetime, effects on light scattering, and deposition in human lungs. However, it is important to note that when we speak of particulate size we are not referring to particulate shape. This is an obvious point, but also a very important one, because for simplicity and for theoretical and learning applications, it is common practice to base assumptions on particles being spherical. Particles are not spherical; they are usually quite irregularly shaped. In pollution control technology design applications, the size of the particle is taken into consideration, but it is the particle's behavior in a gaseous waste stream that we are most concerned with. In order to better understand this important point, consider an analogy. If you take a flat piece of regular typing paper and drop it from the top of a 6-foot ladder, the paper will settle irregularly to the floor, but crumple the same sheet of paper and drop it from the ladder and see what happens. The crumpled paper will fall rapidly. Thus, size is important but other factors, such as particle shape, are also important.

The aerodynamic diameter for all particles greater than 0.5 micrometer can be approximated using the flowing equation. For particles less than 0.5 micrometer, refer to aerosol textbooks to determine the aerodynamic diameter.

$$d_{pa} = d_{ps} \sqrt{P_p} \tag{4.6}$$

where
d_{pa} = aerodynamic particle diameter, μm
d_{ps} = Stokes diameter, μm
P_p = particle density, gm/cm^3

Along with particle size and shape, particle density must be taken into consideration. For example, a baseball and a whiffle ball have approximately the same diameter, but behave quite differently when tossed into the air. Particle density affects the motion of a particle through a fluid and is taken into account in Equation 4.6. The Stokes diameter for a particle is the diameter of the sphere that has the same density and settling velocity as the particle. It is based on the aerodynamic drag force caused by the difference in velocity of the particle and the surrounding fluid. For smooth, spherical particles, the Stokes diameter is identical to the physical or actual diameter.

Inertial sampling devices such as cascade impactors are used for particle sizing. These sampling devices determine the aerodynamic diameter. The term *aerodynamic diameter* is useful for all particles, including fibers and particle clusters. It is not a true size because nonspherical particles require more than one dimension to characterize their size.

Note: In this text, the terms *particle diameter* and *particle size* refer to the aerodynamic diameter unless otherwise stated.

Did You Know?

Particles that appear to have different physical sizes and shapes can have the same aerodynamic diameter. Conversely, some particles that appear to be visually similar can have somewhat different aerodynamic diameters.

PARTICLE SIZE CATEGORIES

Since the range of particle sizes of concern for air emission evaluation is quite broad, it is beneficial to divide this range into smaller categories. Defining different size categories is useful since particles of different sizes behave differently in the atmosphere and the respiratory system.

USEPA has defined four terms for categorizing particles of different sizes. Table 4.2 displays the EPA terminology along with the corresponding particle sizes.

REGULATED PARTICULATE MATTER CATEGORIES

In addition to the terminology provided in Table 4.2, the EPA also categorizes particles as follows:

- *Total suspended particulate matter* (TSP)—Particles ranging in size from 0.1 micrometer to about 30 micrometers in diameter are referred to as total suspended particulate matter (TSP). TSP includes a broad range of particle sizes, including fine, coarse, and supercoarse particles.

TABLE 4.2
EPA Terminology for Particle Sizes

EPA Description	Particle Size
Supercoarse	$d_{pa} > 10 \ \mu m$
Coarse	$2.5 \ \mu m < d_{pa} \leq 10 \ \mu m$
Fine	$0.1 \ \mu m < d_{pa} \leq 2.5 \ \mu m$
Ultrafine	$d_{pa} \leq 0.1 \ \mu m$

- *PM₁₀*—The EPA defines PM_{10} as particulate matter with a diameter of 10 micrometers collected with 50% efficiency by a PM_{10} sampling collection device. However, for convenience in this text, the term *PM_{10}* will be used to include all particles having an aerodynamic diameter of less than or equal to 10 micrometers.

  PM_{10} is regulated as a specific type of pollutant because this size range is considered respirable. In other words, particles less than approximately 10 micrometers can penetrate into the lower respiratory tract. The particle size range between 0.1 and 10 micrometers is especially important in air pollution studies. A major fraction of the particulate matter generated in some industrial sources is in this size range.

- *PM₂.₅*—As with PM_{10}, EPA defines $PM_{2.5}$ as particulate matter with a diameter of 2.5 micrometers with 50% efficiency by a $PM_{2.5}$ sampling collection device. However, for convenience in this text, the term *$PM_{2.5}$* will be used to include all particles having an aerodynamic diameter of less than or equal to 2.5 micrometers.

  The EPA chose 2.5 micrometers as the partition between fine and coarse particulate matter. Particles less than approximately 2.5 micrometers are regulated as $PM_{2.5}$. Air emission testing and air pollution control methods for $PM_{2.5}$ particles are different than those for coarse and supercoarse particles.

  $PM_{2.5}$ particles settle quite slowly in the atmosphere relative to coarse and supercoarse particles. Normal weather patterns can keep $PM_{2.5}$ particles airborne for several hours to several days and enable these particles to cover hundreds of miles. $PM_{2.5}$ particles can cause health problems due to their potentially long airborne retention time and the inability of the human respiratory system to defend itself against particles of this size.

  In addition, the chemical makeup of $PM_{2.5}$ particles is quite different than for coarse and supercoarse particles. EPA data indicate that $PM_{2.5}$ particles are composed primarily of sulfates, nitrates, organic compounds, and ammonium compound. The EPA also determined that $PM_{2.5}$ particles often contain acidic materials, metals, and other contaminants believed to be associated with adverse health effects.

Particles less than 1 micrometer in diameter are termed submicrometer particles and can be the most difficult size to collect. Particles in the range of 0.2 to 0.5 micrometer are common in many types of combustion, waste incineration, and metallurgical sources. Particles in the range of 0.1 to 1.0 micrometer are important because they can represent a significant fraction of the particulate emissions from some types of industrial sources and because they are relatively hard to collect.

Particles less than 0.1 μm—Particles can be much smaller than 0.1 micrometer. In fact, particles composed of as little as 20 to 50 molecules clustered together can exist in a stable form. Some industrial processes such as combustion and metallurgical sources generate particles in the range of 0.01 to 0.1 micrometer. These sizes are approaching the size of individual gas molecules, which are in the range of 0.0002 to 0.001 micrometer. However, particles in the size range of 0.01 to 0.1 micrometer tend to agglomerate rapidly to yield particles in the greater than 0.1 micrometer range. Accordingly, very little of the particulate matter entering an air pollution control device or leaving the stack remains in the very small size range of 0.01 to 0.1 micrometer.

• *Condensable particulate matter*—Particulate matter that forms from condensing gases or vapors is referred to as condensable particulate matter. Condensable particulate matter forms by chemical reactions as well as by physical phenomena.

Condensable particulate matter is usually formed from material that is not particulate matter at stack conditions but which condenses or reacts upon cooling and dilution in the ambient air to form particulate matter. The formation of condensable particulate matter occurs within a few seconds after discharge from the stack.

From a health standpoint, condensable particulate matter is important because it is almost entirely contained in the $PM_{2.5}$ classification.

These particle categories are important because particulate matter is regulated and tested for under them.

SIZE DISTRIBUTION

Particulate emissions from both man-made and natural sources do not consist of particles of any one size. Instead, they are composed of particles over a relatively wide size range. It is often necessary to describe this size range. Particulate matter for size distribution evaluation is measured in a variety of ways. The data must be measured in a manner whereby it can be classified into successive particle diameter size categories.

PARTICLE FORMATION

The range of particle sizes formed in a process is largely dependent on the types of particle formation mechanisms present. The general size range of particles can be estimated by simply recognizing which particle formation mechanisms are

most important in the process being evaluated. The most important particle formation mechanisms in air pollution sources include physical attrition/mechanical
dispersion, combustion particle burnout, homogeneous and heterogeneous nucleation, and droplet evaporation. Several particle formation mechanisms can be present in an air pollution source. As a result, the particles created can have a wide range
of sizes and chemical composition. Particle formation mechanisms are described in
detail below.

- *Physical attrition*—Generates primarily moderate- to large-sized particles and occurs when two surfaces rub together. For example, the grinding
  of a metal rod on a grinding wheel yields small particles that break off
  from both surfaces. The compositions and densities of these particles are
  identical to those of the parent materials.

  In order for fuel to burn, it must be pulverized or atomized so that there
  is sufficient surface area exposed to oxygen and high temperature. As
  indicated in Table 4.3, the surface area of particles increases substantially
  as more and more of the material is reduced in size.

  Accordingly, most industrial-scale combustion processes use one or
  more types of physical attrition in order to prepare or introduce their fuel
  into the furnace. For example, oil-fired boilers use pulverizers to reduce
  the chunks of coal to sizes that can be burned quickly. Oil-fired boilers
  use atomizers to disperse the oil as fine droplets. In both cases, the fuel
  particle size range is reduced to primarily 10 to 1,000 micrometers by
  physical attrition.
- *Combustion particle burnout*—When fuel particles are injected into the hot
  furnace area of the combustion process, such as in fossil-fuel-fired boilers,
  most of the organic compounds in the fuel are vaporized and oxidized in
  the gas stream. Fuel particles become smaller as the volatile matter leaves
  and they are quickly reduced to only the incombustible matter (ash) and the
  slow-burning char composed of organic compounds. Eventually, most of
  the char will also burn, leaving primarily the incombustible material.

TABLE 4.3
Surface Area Comparison for Spherical Particles of Different Diameters[a]

Total Mass	Diameter of Particles (µm)	Number of Particles (Approx. in millions)	Total Surface Area (cm²)	(m²)
	1,000	0.002	60	0.006
	100	2	600	0.06
1.0 g	10	2,000	6,000	0.6
	1	2,000,000	60,000	6
	0.1	2,000,000,000	600,000	60

[a]Based on density of 1.0 g/cm³.

Source: EPA 2007.

As combustion progresses, the fuel particles, which started as 10- to 1,000-micrometer particles, are reduced to ash and char particles that are primarily in the 1- to 100-micrometer range. This mechanism for particle formation can be termed combustion particle burnout.

- *Homogenous and heterogeneous nucleation*—Involve the conversion of vapor phase materials to a particulate form. In both cases, the vapor-containing gas streams must cool to the temperature at which nucleation can occur, which is the dew point. Each vapor phase element and compound has a different dew point. Therefore, some materials nucleate in relatively hot gas zones while others remain as vapor until the gas stream is cold.

Homogeneous nucleation is the formation of new particles composed almost entirely of the vapor phase material. The formation of particles by homogeneous nucleation involves only one compound.

Heterogeneous nucleation is the accumulation of material on the surfaces of existing particles. In the case of heterogeneous nucleation, the resulting particle consists of more than one compound.

There are two main categories of vapor phase material air pollution source gas streams: (1) organic compounds and (2) inorganic metals and metal compounds. In a waste incinerator, waste that volatizes to organic vapor is generally oxidized completely to carbon dioxide and water.

However, if there is an upset in the combustion process, a portion of the organic compounds or their partial oxidation products remain in the gas stream as it leaves the incinerator. Volatile metals and metal compounds such as mercury, lead, lead oxide, cadmium, cadmium oxide, and arsenic trioxide can also volatilize in the hot incinerator. Once the gas stream passes through the heat exchange equipment (i.e., waste heat boiler) used to produce stream, the organic vapors and metal vapors can condense homogeneously or heterogeneously. Generally, the metals and metal compounds search their dew point first and begin to nucleate in relatively hot zones of the unit.

The organic vapors begin to condense in areas downstream from the process, where the gas temperatures are cooler. These particles must then be collected in the downstream air pollution control systems. Homogeneous and heterogeneous nucleation generally create particles that are very small, often between 0.1 and 1.0 micrometer.

Heterogeneous nucleation facilitates a phenomenon called enrichment of particles in the submicrometer size range. The elemental metals and metal compounds volatized during high-temperature operations (fossil fuel combustion, incineration, industrial furnaces, and metallurgical processes) nucleate preferentially as small particles or on the very small particles produced by these processes. Consequently, very small particles have more potentially toxic materials than the very large particles leaving the processes. Heterogeneous nucleation contributes to the formation of article distributions that have quite different chemical compositions in different size ranges.

Another consequence of particle formation by heterogeneous nucleation is that a greater variety of chemical reactions may occur in the gas stream than would otherwise happen. During heterogeneous nucleation, small quantities of metals are deposited on the surfaces of many small particles. In this form, the metals are available to participate in catalytic reactions with gases or other vapor phase materials that are continuing to nucleate. Accordingly, heterogeneous nucleation increases the types of chemical reactions that can occur as the particles travel in the gas stream from the process source and through the air pollution control device.

- *Droplet evaporation*—Some air pollution control systems use solids-containing water recycled from wet scrubbers to cool the gas streams. This practice inadvertently creates another particle formation mechanism that is very similar to fuel burnout. The water streams are atomized during injection into the hot gas streams. As these small droplets evaporate to dryness, the suspended and dissolved solids are released as small particles. The particle size range created by this mechanism has not been extensively studied. However, it probably creates particles that range in size from 0.1 to 20 micrometers. All of these particles must then be collected in the downstream air pollution control systems.

COLLECTION MECHANISMS

When sunlight streams into a quiet room, particles of many different shapes and sizes can be seen; some appear to float while others slowly settle to the floor. All of these small particles are denser than the room air, but they do not settle very fast. The solid, liquid, and fibrous particles formed in air pollution sources behave in a manner that is very similar to standard household dusts and other familiar particles. What we instinctively understand about these everyday particles can be applied in many respects to the particles from air pollution sources.

There are, however, two major differences between industrially generated particles and those in more familiar settings. The industrial particles are much smaller than most household particles. Also, some industrial particles have complex chemical compositions and include compounds and elements that are known to be toxic.

Emission testing devices and air pollution control systems apply forces to the particles in order to remove them from the gas stream. These forces include inertial impaction and interception, Brownian diffusion, gravitational settling, electrostatic attraction, thermophoresis, and diffusiophoresis. These forces are basically the tools that can be used for separating particles from the gas stream. All these collection mechanism forces are strongly dependent on particle size.

- *Inertial impaction and interception*—Due to inertia, a particle moving in a gas stream can strike slowly moving or stationary obstacles (targets) in its path. As the gas stream deflects around the obstacle, the particle

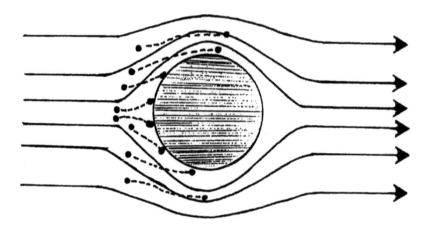

FIGURE 4.1 Particle collection on a stationary object.

continues toward the object and impacts it. The obstacle may be a solid particle, stationary object (as shown in Figure 4.1), or water droplet.

Two primary factors affect the probability of an impaction occurring: (1) aerodynamic article size and (2) the difference in velocity between the particle and the obstacle. Larger particles are collected more easily than smaller particles due to their greater inertia. Also, collection efficiency increases as the difference in velocity between the particle in the gas stream and the obstacle (or target) increases.

Inertial impaction is analogous to a small car riding down an interstate highway at 65 mph and approaching a merge lane where a slowly moving truck is entering. If the car is unable to get into the passing lane to go around the merging truck, there could be an impaction incident. Larger cars will have more difficulty going around the truck than smaller cars. Also, the faster the car is going relative to the truck, the more probable is an impaction.

The efficiency of impaction is directly proportion to the impaction parameter shown in Equation 4.7. As the value of this parameter increases, the efficiency of inertial impaction increases. This parameter is related to the square of the Stokes particle diameter and the difference in velocity between the particle and the target droplet.

$$K_1 = \frac{C_c (d_{ps})^2 P_p}{18 \mu D_c} \qquad (4.7)$$

where
K_1 = impaction parameter (dimensionless)
C_c = Cunningham slip correction factor (dimensionless)
d_{ps} = Stokes particle diameter (micrometers)
v = difference in velocity (cm/s)

P_p= particle density (gm/cm³)
D_c= diameter of a droplet (cm)
μ = gas viscosity (gm/cm-s)

The Cunningham slip correction factor (also called Cunningham's correction factor) accounts for molecular slip. Molecular slip occurs when the size of the particle is in the same magnitude as the distance between gas molecules. The particle no longer moves as a continuum in the gas, but as a particle among discrete gas molecules, thereby reducing the drag force. For particles in air with an actual diameter of 1.0 micrometer or less, the Cunningham correction factor is significant.

Inertial impaction occurs when obstacles (e.g., water droplets) are directly in the path of the particle moving in the gas stream. Sometimes the obstacle or target is offset slightly from the direct path of the moving particle. In this instance, as the particle approaches the edge of the obstacle, the obstacle may collect the particle through a process called *interception*.

Did You Know?

Inertial impaction and interception are usually highly efficient for particles larger than 10 micrometers. They become progressively less effective as the size decreases. Impaction is not efficient for particles less than 0.3 micrometers due to their low inertia.

- *Brownian diffusion*—Becomes the dominant collection mechanism for particles less than 0.3 micrometer and is especially significant for particles in the 0.01- to 0.1-micrometer size range.

Very small particles in a gas stream deflect slightly when gas molecules strike them. Transfer of kinetic energy from the rapidly moving gas molecule to the small particle causes this deflection, called Brownian diffusion. These small particles are captured when they impact a target (e.g., liquid droplet) as a result of this random movement.

Diffusivity is a measure of the extent to which molecular collisions influence very small particles, causing them to move in a random manner across the direction of gas flow. The diffusion coefficient in the equation below represents the diffusivity of a particle at certain gas stream conditions.

$$D_p = \frac{C_c K T}{3\pi d_{pa}\mu} \tag{4.8}$$

where
D_p= diffusion coefficient (cm²/s)
C_c = Cunningham slip correction factor (dimensionless)

K = Boltzmann constant (gm-cm^2/s^2 K)
T = absolute temperature (K)
d_{pa} = particle aerodynamic diameter (micrometers)
μ = gas viscosity (kg/m-s)

- *Gravitational settling*—Particles in still air have two forces acting on them: (1) a gravitational force downward and (2) the air resistance (or drag) force upward. When particles begin to fall, they quickly reach a terminal settling velocity, which represents the constant velocity of a falling particle when the gravitational force downward is balanced by the air resistance (or drag) force upward. The terminal settling velocity can usually be expressed using Equation 4.9. (Note: Equation 4.9 is applicable for particles less than 80 micrometers in size (aerodynamic diameter) and having a Reynolds number less than 2.0 and a low velocity.)

$$v_t = \frac{GP_p(d_{ps})^2 C_c}{18\,\mu} \tag{4.9}$$

where
v_t = terminal settling velocity (cm/s)
g = gravitational acceleration (cm/s^2)
P_p = density of particle (gm/cm^3)
d_{ps} = Stokes particle diameter (cm)
C_c = Cunningham slip correction factor (dimensionless)
μ = viscosity of air (gm/cm-s)

Did You Know?

Due to the low settling velocities of essentially all particles less than 80 micrometers (aerodynamic diameter), gravitational settling of particles is not used for the initial separation of particles from the gas stream. However, this does not mean that gravitational settling is unimportant. Settling by gravity plays an important role in the removal of large clumps of dust when fabric filter bags and electrostatic precipitation collection plates are cleaned.

- *Electrostatic attraction*—In air pollution control, electrostatic precipitators (ESPs) use electrostatic attraction for particulate collection. Electrostatic attraction of particles is accomplished by establishing a strong electrical field and creating unipolar ions. The particles passing through the electrical field are charged by the ions being driven along the electrical field lines. Several parameters dictate the effectiveness of electrostatic attraction, including the particle size, gas flow rate, and resistivity.
  The particles will eventually reach a maximum or saturation charge, which is a function of the particle area. The saturation charge occurs when the localized field created by the already captured ions is

sufficiently strong to deflect the electrical field lines. Particles can also be charged by diffusion of ions in the gas stream. The strength of the electrical charges imposed on the particles by both mechanisms is particle size dependent.

Did You Know?

Electrostatic precipitators are designed based on the flow rate of the flue gas to be treated. Flow rates that exceed design specifications may result in increased penetration of particulate matter. Flow rates that are too low may allow particulate matter to drop out of the gas stream near the inlet of the ESP and lead to buildup of particulated matter in the ducts.

Resistivity is a measure of the ability of the particle to conduct electricity and is expressed in units of ohm-cm. Particles with low resistivity have a greater ability to conduct electricity (and higher electrostatic attraction) than particles with high resistivity. The following factors influence resistivity:
   • Chemical composition of the gas stream
   • Chemical composition of the particle
   • Gas stream temperature
 • *Thermophoresis*—Particle movement caused by thermal differences on two sides of the particle. Gas molecules at higher temperatures have greater kinetic energy than those at lower temperatures. Therefore, when the particle collides with a gas molecule from the hotter side, the particle receives more kinetic energy than when it collides with a gas molecule from the cooler side. Accordingly, particles tend to be deflected toward the colder area.
 • *Diffusiophoresis*—Particle movement caused by concentration differences on two sides of the particle. When there is a strong difference in the concentration of gas molecules on two sides of the particle, there is a difference in the number of molecular collisions, which causes an imbalance in the total kinetic energies of the gas molecules. Gas molecules in the high-concentration area striking a particle transmit more kinetic energy to the particle than molecules in the lower-concentration area. Therefore, particles tend to move toward the area of lower concentration.

Did You Know?

Diffusiophoresis can be important when the evaporation or condensation of water is involved since these conditions create substantial concentration gradients. The normal differences in pollutant concentration are not sufficient to cause significant particle movement.

REFERENCES AND RECOMMENDED READING

EPA. 2007. Basic concepts in environmental sciences. www.epa.gov/eogapti1/module3/col-
    lect/collect.htm (accessed January 2, 2008).
Hinds, W. C. 1982. *Aerosol technology, properties, behavior, and measurement of airborne
    particles*. New York: Wiley-Interscience Publication.
Reist, P. C. 1982. *Introduction to aerosol science*. New York: Macmillian.
Spellman, F. R., and Whiting, N. 2006. *Environmental science and technology: Concepts
    and applications*. 2nd ed. Rockville, MD: Government Institutes.

5 Basic Air Chemistry

We are normally too engrossed with the activities of our daily lives to consider what life would be like without chemistry. If we were to ponder this point, however, it would soon be apparent that chemistry affects everything that we do. Not a single moment of time goes by which we are not affected somehow by a chemical substance or chemical process.

Eugene Meyer (1989)

INTRODUCTION

Chemistry, chemists, chemicals? As Meyer points out above, we normally give little thought to these terms. When we do, what do we think? Some folks immediately tense up when the term *chemistry* is mentioned. To them, chemistry is a strange term with stranger connotations.

There are many views of chemistry and chemists (in particular)—too many different perceptions to describe here. One thing is certain, however. When we mention chemistry to almost anyone, the image that almost always forms is one of the mixing of liquid chemicals. People have the perception that chemistry is most often something that happens in liquids, especially the exotic liquids—the colored ones that are featured in horror stories, science fiction tales, and science classes. Is it any surprise that the average person would never equate chemistry to something in the atmosphere as well? They do not know about chemistry to understand the multitude of reactions—slow and fast chemical reactions, dissolving chemicals, precipitation of colored solids—that occur in air.

Countless experiments have been conducted in earth's atmosphere over that past few centuries. Consider, for example, the weather balloons that we have sent up with special measuring instruments. The instruments measure certain atmospheric conditions to aid us in forecasting the weather. These balloons have done much more for us than that. They have revealed that the atmosphere contains several thousand different chemical species.

Experiments have also been conducted by satellites circling the globe, and some airplanes have been virtually outfitted as flying chemical laboratories. Experiments have been conducted on the earth's surface as well, all of which have also helped to catalog thousands of different chemical species. In chemistry, and in science in general, we are always making new discoveries—always trying to find out something about the "we do not know what we do no know" syndrome.

But we know a lot. For example, as Graedel and Crutzen (1995) put it, we know that earth's atmosphere is literally a "flask without walls," containing several thousand different chemical species, many of which consist of the atmospheric gas itself. Some are found in airborne particulate matter, both large and small, and some are found dissolved in hydrometeors. Many atmospheric compounds have both natural and anthropogenic origins. The compounds from natural origins (volcanic action, earthquakes, horrendous storms, and others) have been with us from the very beginning. Anthropogenic contributions are more recent—from the industrial revolution up through today.

Why are we concerned with air chemistry? Why study air chemistry at all? In a general sense, consider that chemistry basically affects everything that we do. Not a single moment of time goes by during which we are not affected in some way by a chemical substance, chemical process, or chemical reaction. Chemistry affects just about every area of our daily lives. In a specific sense, consider that almost every environmental and pollution problem we face today (and probably tomorrow) has a chemical basis. Without chemistry, it would be difficult, if not impossible, to study problems related to the quality of air that living organisms breathe, the nature and level of air pollutants, visibility, atmospheric esthetics, and climate, as well as those phenomena affecting the atmosphere (greenhouse effect, ozone depletion, groundwater contamination, toxic wastes, air pollution, and acid rain).

The task of the air science practitioner is to understand at least the most influential of the reactions linking atmospheric chemical constituents, as well as their principal sources and removal technologies. A basic review of chemistry fundamentals, chemical reactions in the atmosphere, and the chemical nature of atmospheric chemical species is given in this chapter.

Note that the air science practitioner is concerned in general with the atmospheric cycle—the same general phenomenon that governs and produces all aspects of atmospheric chemistry. We will discuss air science as it relates to the unpolluted atmosphere, to highly polluted atmospheres, and to a wide range of gradations in between.

CHEMISTRY FUNDAMENTALS

This text assumes that the air science student or interested reader who uses this book may or may not have some fundamental knowledge of chemistry. Thus, in this section, the topics have been selected with the goal of reviewing or providing only the essential chemical principles required to understand the nature of air science problems that we face and the chemistry involved with scientific and technological approaches to their solutions.

WHAT IS CHEMISTRY?

Chemistry is the science concerned with the composition of matter (gas, liquid, or solid) and of the changes that take place in it under certain conditions.

Every substance, material, and object in the environment is either a chemical substance or mixture of chemical substances. Your body is made up of chemicals, literally thousands of them. The food we eat, the clothes we wear, the fuel we burn, and the vitamins we take in from natural or synthetic sources are all products of chemistry, wrought by either the forces of nature or the hand of man. Chemistry is about matter—its constituents and consistency—and about measuring and quantifying matter.

What is matter? All matter can exist in three states: gas, liquid, or solid. Matter is composed of minute particles termed molecules, which may be further divided into atoms. Molecules that contain atoms of one kind only are known as *elements*; those that contain atoms of different kinds are called *compounds*.

Chemical compounds are produced by a chemical action that alters the arrangements of the atoms in the reacting molecules. Heat, light, vibration, catalytic action, radiation, or pressure, as well as moisture (for ionization), may be necessary to produce a chemical change. Examination and possible breakdown of compounds to determine their components is *analysis*, and the building up of compounds from their components is *synthesis*. When substances are brought together without changing their molecular structures they are said to be *mixtures*.

Organic substances consist of virtually all compounds that contain carbon. All other substances are *inorganic substances*.

ELEMENTS AND COMPOUNDS

A pure substance is a material that has been separated from all other materials. Examples of such substances (indistinguishable from pure samples of the substance, no matter what procedures are used to purify them or what their origin is) are copper metal, aluminum metal, distilled water, table sugar, and oxygen. All samples of table sugar are alike and indistinguishable from all other samples.

Usually expressed in terms of percentage by mass, a substance is characterized as a material having a fixed composition. Distilled water, for example, is a pure substance consisting of approximately 11% hydrogen and 89% oxygen by mass. By contrast, a lump of coal is not a pure substance; it is a mixture, and its carbon content may vary from 35% to 90% by mass.

When substances cannot be broken down or decomposed into simpler forms of matter, they are called elements. The elements are the basic substances of which all matter is composed. At the present time there are only 100+ known elements but well over a million known compounds. Of the 100+ elements, only 88 are present in detectable amounts on earth, and many are rare. Table 5.1 shows that only ten elements make up approximately 99% by mass of the earth's crust, including the

TABLE 5.1
Elements Making up 99% of Earth's Crust, Oceans, and Atmosphere

Element	Symbol	% of Composition	Atomic number
Oxygen	O	49.5%	8
Silicon	Si	25.7%	14
Aluminum	Al	7.5%	13
Iron	Fe	4.7%	26
Calcium	Ca	3.4%	20
Sodium	Na	2.6%	11
Potassium	K	2.4%	19
Magnesium	Mg	1.9%	12
Hydrogen	H	1.9%	1
Titanium	Ti	0.58%	22

surface layer, the atmosphere, and the bodies of water. From Table 5.1, we see that the most abundant element on earth is oxygen, which is found in the free state in the atmosphere, as well as in combined form with other elements in numerous minerals and ores. Table 5.1 also provides the symbols and atomic numbers of the ten chemicals listed. The symbols consist of either one or two letters, with the first letter capitalized. The *atomic number* of an element is the number of protons in the nucleus.

Classification of Elements

Each element may be classified as a metal, nonmetal, or metalloid. *Metals* are typically lustrous solids that are good conductors of heat and electricity; they melt and boil at high temperatures, possess relatively high densities, and are normally malleable (can be hammered into sheets) and ductile (can be drawn into a wire). Examples of metals are copper, iron, silver, and platinum. Almost all metals are solids (none is gaseous) at room temperature (mercury being the only exception).

Elements that do not possess the general physical properties just mentioned (i.e., they are poor conductors of heat and electricity, boil at relatively low temperatures, do not possess a luster, and are less dense than metals) are called *nonmetals*. Most nonmetals at room temperature are either solids or gases (the exception is bromine— a liquid). Nitrogen, oxygen, and fluorine are examples of gaseous nonmetals, while sulfur, carbon, and phosphorus are examples of solid nonmetals.

Several elements have properties resembling both metals and nonmetals. They are called *metalloids* (semimetal). The metalloids are boron, silicon, germanium, arsenic, tellurium, antimony, and polonium.

Physical and Chemical Changes

Internal linkages among a substance's units (between one atom and another) maintain its constant composition. These linkages are called chemical bonds. When a particular process occurs that involves the making and breaking of these bonds, we say that a chemical change or chemical reaction has occurred. In environmental science, combustion and corrosion are common examples of chemical changes that impact our environment.

Let us briefly consider a couple examples of chemical change. When a flame is brought into contact with a mixture of hydrogen and oxygen gases, a violent reaction takes place. The covalent bonds in the hydrogen (H_2) molecules and of the oxygen (O_2) moles are broken and new bonds are formed to produce molecules of water, H_2O.

The key point to remember is that whenever chemical bonds are broken or formed, or both, a chemical change takes place. The hydrogen and oxygen undergo a chemical change to produce water, a substance with new properties.

When mercuric oxide (a red powder) is heated, small globules of mercury are formed and oxygen gas is released. This mercuric oxide is changed chemically to form molecules of mercury and water.

By contrast, a physical change (nonmolecular change) is one in which the molecular structure of a substance is not altered. When a substance freezes, melts, or changes to vapor, the composition of each molecule does not change. For example, ice, steam, and liquid water all are made up of molecules containing two atoms of hydrogen and one atom of oxygen. A substance can be ripped or sawed into small pieces, ground into powder, or molded into a different shape without changing the molecules in any way.

The types of behavior that a substance exhibits when undergoing chemical changes are called its chemical properties. The characteristics that do not involve changes in the chemical identity of a substance are called its physical properties. All substances may be distinguished from one another by these properties, in much the same way as certain features (DNA, for example) distinguish one human being from another.

THE STRUCTURE OF THE ATOM

If a small piece of an element (say, copper) is hypothetically divided and subdivided into the smallest piece possible, the result would be one particle of copper. This smallest particle of the element, which is still representative of the element, is called an *atom*.

Although infinitesimally small, the atom is composed of particles, principally electrons, protons, and neutrons. The simplest atom possible consists of a nucleus with a single *proton* (positively charged particles) and a single *electron* (negatively charged particles) traveling around it—an atom of hydrogen, which has an atomic weight of 1 because of the single proton. The atomic weight of an element is equal to the total number of protons and neutrons (neutral particles) in the nucleus of an atom of an element. Electrons and protons bear the same magnitude of charge, but of opposite polarity.

The hydrogen atom also has an atomic number of 1 because of its one proton. The *atomic number* of an element is equal to the number of protons in its nucleus. A neutral atom has the same number of protons and electrons. Therefore, in a neutral atom the atomic number is also equal to the number of electrons in the atom. The number of neutrons in an atom is always equal to or greater than the number of protons except in the atom of hydrogen.

The protons and neutrons of an atom reside in the nucleus. Electrons reside primarily in designated regions of space surrounding the nucleus, called *atomic orbitals* or *electron shells*. Only a prescribed number of electrons may reside in a given type of electron shell. With the exception of hydrogen (which only has one electron), two electrons are always close to the nucleus, in an atom's innermost electron shell. In most atoms, other electrons are located in electron shells some distance from the nucleus.

While neutral atoms of the same element have an identical number of electrons and protons, they may differ by the number of neutrons in their nuclei. Atoms of the same element having different numbers of neutrons are called *isotopes* of that element.

PERIODIC CLASSIFICATION OF THE ELEMENTS

Through experience, scientists discovered that the chemical properties of the elements repeat themselves. Chemists summarize all such observations in the *periodic law*: the properties of the elements vary periodically with their atomic numbers.

In 1869, Dimitri Mendeleev, using relative atomic masses, developed the original form of what today is known as the *periodic table*—a chart of elements arranged in

11	Atomic Number
Sodium	Name
Na	Symbol

FIGURE 5.1 The element sodium as it is commonly shown in one of the horizontal boxes in the periodic table.

order of increasing proton number to show the similarities of chemical elements with related electronic configurations. The elements fall into vertical columns, known as *groups*. Going down a group, the atoms of the elements all have the same outer shell structure but an increasing number of inner shells. Traditionally, the alkali metals are shown on the left of the table and the groups are numbered IA to VIIA, IB to VIIB, and 0 (for noble gases). It is now more common to classify all the elements in the middle of the table as transition elements and to regard the nontransition elements as main-group elements numbered from I to VII, with the noble gases in group 0. Horizontal rows in the table are *periods*. The first three are called short periods; the next four (which include transition elements) are long periods. Within a period, the atoms of all the elements have the same number of shells, but with a steadily increasing number of electrons in the outer shell.

The periodic table is an important tool for learning chemistry because it tabulates a variety of information in one spot. For example, we can immediately determine the atomic number of the elements because they are tabulated on the periodic table (see Figure 5.1). We can also readily identify which elements are metals, nonmetals, and metalloids. Usually a bold zigzag line separates metals from nonmetals, while those elements lying to each immediate side of the line are metalloids. Metals fall to the left of the line, and nonmetals fall to the right of it.

MOLECULES AND IONS

When elements other than noble gases (which exist as single atoms) exist in either the gaseous or liquid state of matter at room conditions, they consist of units containing pairs of like atoms. These units are called *molecules*. For example, we generally encounter oxygen, hydrogen, chlorine, and nitrogen as gases. Each exists as a molecule having two atoms. These molecules are symbolized by the notations O_2, H_2, Cl_2, and N_2, respectively.

The smallest particle of many compounds is also the molecule. Molecules of compounds contain atoms of two or more elements. The water molecule, for example, consists of two atoms of hydrogen and one atom of oxygen (H_2O). Methane molecules consist of one carbon atom and four hydrogen atoms (CH_4).

Not all compounds occur naturally as molecules. Many of them occur as aggregates of oppositely charged atoms (or groups of atoms) called *ions*. Atoms become charged by gaining or losing some of their electrons. Atoms of metals, for example, that lose their electrons become positively charged, and atoms of nonmetals that gain electrons become negatively charged.

CHEMICAL BONDING

When compounds form, the atoms of one element become attached to, or associated with, atoms of other elements by forces called chemical bonds. Chemical bonding is a strong force of attraction holding atoms together in a molecule. There are various types of chemical bonds. *Ionic* bonds can be formed by transfer of electrons. For instance, the calcium atom has an electron ion figuration of two electrons in its outer shell. The chlorine atom has seven outer electrons. If the calcium atom transfers two electrons (one to each chlorine atom), it becomes a calcium ion with the stable configuration of an inert gas. At the same time, each chlorine (having gained one electron) becomes a chlorine ion, also with an inert gas configuration. The bonding in calcium chloride is the electrostatic attraction between the ions.

Covalent bonds are formed by sharing of *valence* (the number of electrons an atom can give up or acquire to achieve a filled outer shell) electrons. Hydrogen atoms, for instance, have one outer electron. In the hydrogen molecule, H_2, each atom contributes one electron to the bond. Consequently, each hydrogen atom has control of two electrons—one of its own and the second from the other atom—giving it the electron configuration of an inert gas. In the water molecule (H_2O), the oxygen atom (with six outer electrons) gains control of an extra two electrons supplied by the two hydrogen atoms. Similarly, each hydrogen atom gains control of an extra electron from the oxygen.

Chemical compounds are often classified into either of two groups based on the nature of the bonding between their atoms. And, as you might expect, chemical compounds consisting of atoms bonded together by means of ionic bonds are called ionic compounds. Compounds whose atoms are bonded together by covalent bonds are called covalent compounds.

Most ionic and covalent compounds provide some interesting contrasts. For example, ionic compounds have higher melting points, boiling points, and solubility in water than covalent compounds. Ionic compounds are nonflammable while covalent compounds are flammable. Ionic compounds molten in water solutions conduct electricity. Molten covalent compounds do not conduct electricity. Ionic compounds generally exist as solids at room temperature, while covalent compounds exist as gases, liquids, and solids at room temperature.

CHEMICAL FORMULAS AND EQUATIONS

Chemists have developed a shorthand method of writing chemical formulas. Elements are represented by groups of symbols called formulas. A common compound is sulfuric acid; its formula is H_2SO_4. The formula indicates that the acid is composed of two atoms of hydrogen, one atom of sulfur, and four atoms of oxygen. However, this is not a recipe for making the acid. The formula does not tell you how to prepare the acid, only what is in the acid.

A *chemical equation* tells what elements and compounds are present before and after a chemical reaction. Sulfuric acid poured over zinc will cause the release of hydrogen and the formation of zinc sulfate. This is shown by the following equation:

$$Zn + H_2SO_4 \quad \rightarrow \quad ZnSO_4 + H_2$$
$$\text{(Zinc)} \quad \text{(Sulfuric acid)} \qquad \text{(Zinc)} \qquad \text{(Hydrogen)}$$

One atom (also one molecule) of zinc unites with one molecule of sulfuric acid, giving one molecule of zinc sulfate and one molecule (two atoms) of hydrogen. Notice that there is the same number of atoms of each element on each side of the arrow. However, the atoms are combined differently.

MOLECULAR WEIGHTS, FORMULAS, AND THE MOLE

The relative weight of a compound that occurs as molecules is called the *molecular weight*—the sum of the atomic weights of each atom that comprises the molecule. Consider the water molecule. Its molecular weight is determined as follows:

$$2 \text{ hydrogen atoms} = 2 \times 1.008 = 2.016$$
$$\underline{1 \text{ oxygen atom} = 1 \times 15.999 = 15.999}$$
$$\text{Molecular weight of } H_2O = 18.015$$

Thus, the molecular weight of a molecule is simply the sum of the atomic weights of all of the constituent atoms. If we divide the mass of a substance by its molecular weight, the result is the mass expressed in *moles* (mol). Usually, the mass is expressed in grams, in which case the moles are written as g-moles; similarly, if mass is expressed in pounds, the result would be lb-moles. One g-mole contains 6.022×10^{23} molecules (*Avogadro's number*, in honor of the scientist who first suggested its existence), and 1 lb-mole about 2.7×10^{26} molecules.

$$\text{Moles} = \frac{\text{Mass}}{\text{Molecular weight}} \qquad (5.1)$$

The relative weight of a compound that occurs as formula units is called the *formula weight*—the sum of the atomic weights of all atoms that comprise one formula unit. Consider sodium fluoride. Its formula weight is determined as follows:

$$1 \text{ sodium ion} = 22.990$$
$$\underline{1 \text{ fluoride ion} = 18.998}$$
$$\text{Formula weight of NaF} = 41.988$$

PHYSICAL AND CHEMICAL PROPERTIES OF MATTER

There are two basic types of properties (characteristics) of matter: physical and chemical. *Physical properties* of matter are those that do not involve a change in the chemical composition of the substance. Among these properties are hardness, color, boiling point, electrical conductivity, thermal conductivity, specific heat, density, solubility, and melting point. These properties may change with a change in temperature or pressure. Those changes that do not alter chemical composition of the substance are called physical changes. When heat is applied to solid ice to convert it to liquid (water) no new substance is produced, but the appearance has changed—melting is a physical change. Other examples of physical changes are dissolving sugar in water, heating a piece of metal, and evaporating water.

The physical properties that are most commonly used in describing and identifying particular kinds of matter are density, color, and solubility. *Density* (d) is mass per unit volume and is expressed by the equation

$$d = \frac{\text{Mass}}{\text{Volume}} \tag{5.2}$$

All matter has weight and takes up space; it also has density, which depends on weight and space. We commonly say that a certain material will not float in water because it is heavier than water. What we really mean is that a particular material is more dense than water. The density of an element differs from the density of any other element. The densities of liquids and solids are normally given in units of grams per cubic centimeter (g/cm^3), which is the same as grams per milliliter (g/ml). The advantage of using the physical property color is that no chemical or physical tests are required. *Solubility* refers to the degree to which a substance dissolves in a liquid such as water. In air science, the density, color, and solubility of a substance are important physical properties that aid in the determination of various pollutants, stages of pollution or treatment, the remedial actions required to clean up volatile toxic/hazardous waste spills, and other environmental problems.

The properties involved in the transformation of one substance into another are known as *chemical properties*. For example, when a piece of wood burns, oxygen in the air unites with the different substances in the wood to form new substances. Another example is corroded iron. During the iron corrosion process, oxygen combines with the iron and water to form a new substance commonly known as rust. Changes that result in the formation of new substances are known as chemical changes.

STATES OF MATTER

As pointed out earlier, the three common states (or phases) of matter are the solid state, the liquid state, and the gaseous state. In the solid state, the molecules or atoms are in a relatively fixed position. The molecules are vibrating rapidly but about a fixed point. Because of this definite position of the molecules, a solid holds its shape. A *solid* occupies a definite amount of space and has a fixed shape.

When the temperature of a gas is lowered, the molecules of the gas slow down. If the gas is cooled sufficiently, the molecules slow down so much that they lose the energy needed to move rapidly throughout their container. The gas may turn into liquid. Common liquids are water, oil, and gasoline. A *liquid* is a material that occupies a definite amount of space, but which takes the shape of the container.

In some materials, the atoms or molecules have no special arrangement at all. Such materials are called gases. Oxygen, carbon dioxide, and nitrogen are common gases. A *gas* is a material that takes the exact volume and shape of its container.

Although the three states of matter discussed above are familiar to most students and others, it is the change from one state to another that is of primary interest to environmentalists. Changes in matter that include water vapor changing from the gaseous state to liquid precipitation, or a spilled liquid chemical changed to a semisolid substance (by addition of chemicals, which aids in the cleanup effort) are just two ways changing from one state to another has an impact on environmental concerns.

THE GAS LAWS

The gas laws were explained earlier in the text; for the purpose of review and because of their importance, they will be explained again briefly.

The atmosphere is composed of a mixture of gases, the most abundant of which are nitrogen, oxygen, argon, carbon dioxide, and water vapor. The *pressure* of a gas is the force that the moving gas molecules exert upon a unit area. A common unit of pressure is newton per square meter, N/m_2, called a pascal (Pa). An important relationship exists among the pressure, volume, and temperature of a gas. This relation is known as the *ideal gas law* and can be stated as

$$\frac{P_1 V_1}{T_1} = \frac{P_2 V_2}{T_2} \tag{5.3}$$

where P_1, V_1, and T_1 are pressure, volume, and absolute temperature at time 1, and P_2, V_2, and T_2, are pressure, volume, and absolute temperature at time 2. A gas is called perfect (or ideal) when it obeys this law.

A temperature of 0°C (273 K) and a pressure of 1 atmosphere (atm) have been chosen as standard temperature and pressure (STP). At STP the volume of 1 mole of ideal gas is 22.4 L.

LIQUIDS AND SOLUTIONS

Solutions, which are homogenous mixtures, can be solid, gaseous, or liquid. However, the most common solutions are liquids. The substance in excess in a solution is called the solvent. The substance dissolved is the solute. Solutions in which water is the solvent are called aqueous solutions. A solution in which the solute is present in only a small amount is called a dilute solution. If the solute is present in large amounts, the solution is a concentrated one. When the maximum amount of solute possible is dissolved in the solvent, the solution is saturated.

The concentration (or amount of solute dissolved) is frequently expressed in terms of the molar concentration. The *molar concentration*, or *molarity*, is the number of moles of solute per liter of solution. Thus, a molar solution (written 1.0 M) has 1 gram formula weight of solute dissolved in 1 liter of solution. In general,

$$\text{Molarity} = \frac{\text{Moles of solute}}{\text{Number of liters of solution}} \tag{5.4}$$

Note that the number of liters of solution (not the number of liters of solvent) is used.

EXAMPLE 5.1

Exactly 40 g of sodium chloride (NaCl; table salt) was dissolved in water, and the solution was made up to a volume 0.80 liter. What was the molar concentration (M) of sodium chloride in the resulting solution?

Solution
First find the number of moles of salt.

$$\text{number of moles} = \frac{40 \text{ g}}{58.5 \text{ g/mole}} = 0.68 \text{ mole}$$

$$\text{molarity} = \frac{0.68 \text{ mole}}{0.80 \text{ L}} = 0.85 \text{ M}$$

THERMAL PROPERTIES

Knowing the thermal properties of chemicals and other substances is important to the air science practitioner. Such knowledge is used in hazardous materials spill mitigation and in solving many other complex environmental problems. Heat is a form of energy. Whenever work is performed, there is usually a substantial amount of heating caused by friction. The conservation of energy law tells us that the work done plus the heat energy produced must equal the original amount of energy available; that is,

$$\text{Total energy} = \text{work done} + \text{heat produced} \qquad (5.5)$$

As air or environmental scientists, technicians, or practitioners, we are concerned with several properties related to heat for particular substances. Those thermal properties that we need to be familiar with are discussed in the following sections.

A traditional unit for measuring heat energy is the calorie. A *calorie* (cal) is defined as the amount of heat necessary to raise 1 gram of pure liquid water by 1°C at normal atmospheric pressure. In SI units, 1 cal = 4.186 J (joule).

Do not confuse the calorie just defined with the one used when discussing diets and nutrition. A kilocalorie is 1,000 calories—the amount of heat necessary to raise the temperature of 1 kilogram of water by 1°C.

In the British system of units, the unit of heat is the British thermal unit, or Btu. One *Btu* is the amount of heat required to raise 1 pound of water 1°F at normal atmospheric pressure (1 atm).

SPECIFIC HEAT

Earlier, it was pointed out that 1 kilocalorie of heat is necessary to raise the temperature of 1 kilogram of water 1°C. Other substances require different amounts of heat to raise the temperature of 1 kilogram of the substance 1 degree. The specific heat of a substance is the amount of heat in kilocalories necessary to raise the temperature of 1 kilogram of the substance 1°C.

The units of specific heat are kcal/kg°C or, in SI units, J/kg°C. The specific heat of pure water, for example, is 1.000 kcal/kg°C. This is 4,186 J/kg°C.

The greater the specific heat of a material, the more heat that is required. The greater the mass of the material or the greater the temperature change desired, the more heat that is required.

The amount of heat necessary to change 1 kilogram of a solid into a liquid at the same temperature is called the *latent heat of fusion* of the substance. The temperature of the substance at which this change from solid to liquid takes place is known as the *melting point*. The amount of heat necessary to change 1 kilogram of a liquid into a gas is called the *latent heat of vaporization*. When this point has been reached, the substance is all in the gas state. The temperature of the substance at which this change from liquid to gas occurs is known as the *boiling point*.

ACID + BASES → SALTS

When acids and bases are combined in the proper proportions they neutralize each other, each losing its characteristic properties and forming a salt and water. For example,

$$NaOH + HCl \rightarrow NaCl + H_2O$$

which is

sodium hydroxide + hydrochloric acid → sodium chloride + water

The acid-base-salt concept originated with the beginning of chemistry and is very important in the environment, the life processes, and in industrial chemistry.

The word *acid* is derived from the Latin *acidus*, which means "sour." The sour taste is one of the properties of acids (however, the student should never actually taste an acid). An *acid* is a substance that, in water, produces hydrogen ions, H_+, and has the following properties:

1. Conducts electricity
2. Tastes sour
3. Changes the color of blue litmus paper to red
4. Reacts with a base to neutralize its properties
5. Reacts with metals to liberate hydrogen gas

A *base* is a substance that produces hydroxide ions, OH^-, or accepts H^+, and when dissolved in water it has the following properties:

1. Conducts electricity
2. Changes the color of red litmus paper to blue
3. Tastes bitter and feels slippery
4. Reacts with an acid to neutralize the acid's properties

PH SCALE

A common way to determine whether a solution is an acid or a base is to measure the concentration of hydrogen ions (H^+) in the solution. The concentration can be expressed in powers of 10 but is more conveniently expressed as *pH* (the *p* is from the German word *poentz* and the *H* stands for *hydrogen*). For example, pure water has

---

TABLE 5.2
pH Scale

pH	Concentration of H-Ions (mole/liter)	Acidic/Basic
1	1.0×10^{-1}	Very acidic
2	1.0×10^{-2}	
3	1.0×10^{-3}	
4	1.0×10^{-4}	
5	1.0×10^{-5}	
6	1.0×10^{-6}	Acidic
7	1.0×10^{-7}	Neutral
8	1.0×10^{-8}	Basic
9	1.0×10^{-9}	
10	1.0×10^{-10}	
11	1.0×10^{-11}	
12	1.0×10^{-12}	
13	1.0×10^{-13}	
14	1.0×10^{-14}	Very basic

1×10^{-7} grams of hydrogen ions per liter. The negative exponent of the hydrogen ion concentration is called the pH of the solution. The pH of water is 7—a neutral solution. A concentration of 1×10^{-12} has a pH of 12. A pH less than 7 indicates an acid solution, and a pH greater than 7 indicates a basic solution (see Table 5.2).

The pH values of substances found in our environment vary. Acid-base reactions are among the most important in environmental science. In diagnosing various environmental problems such as acid rain problems, hazardous materials spills into lakes or streams, and the effect of point or nonpoint source pollution into our streams and rivers, lakes and ponds, pH value is important. Along with diagnosis, remediation and prevention are important. For example, to protect a local ecosystem, volatile wastes often require neutralization before being released into the environment. Table 5.3 gives an approximate pH of some common substances.

ORGANIC CHEMISTRY

Organic chemistry is the branch of chemistry that is concerned with compounds of carbon. The science of organic chemistry is incredibly complex and varied. There are millions of different organic compounds known today and 100,000+ of these are products of synthesis, unknown in nature. The next few pages will provide a very basic introduction to some of the most common and important organic substances that are important to environmental science (important because of their toxicities as pollutants and other hazards), so that they will be less foreign when we encounter them later in the text.

Before 1828, scientists thought that organic compounds could only be made by plants and animals (living things). In that year, Friedrich Wohler made urea from ammonium cyanate, thus disproving the theory that stated urea could only be made from living things. Because of his discovery, the science of organic chemistry was born.

TABLE 5.3
pH of Common Substances

Substance	pH
Battery acid	0.0
Gastric juice	1.2
Lemons	2.3
Vinegar	2.8
Soft drinks	3.0
Apples	3.1
Grapefruit	3.1
Wines	3.2
Oranges	3.5
Tomatoes	4.2
Beer	4.5
Bananas	4.6
Carrots	5.0
Potatoes	5.8
Coffee	6.0
Milk (cow)	6.5
Pure water (Neutral)	7.0
Blood (human)	7.4
Eggs	7.8
Sea water	8.5
Milk of magnesia	10.5
Oven cleaner	13.0

Organic compounds are components of all the familiar commodities that our technological world requires. Examples of such commodities are motor and heating fuels, adhesives, cleaning solvents, paints, varnishes, plastics, refrigerants, aerosols, textiles, fibers, and resins, among many others.

From an environmental science perspective, the principal concern about organic compounds is that they are pollutants of water, air, and soil environments. As such, they are safety and health hazards. They are also combustible or flammable substances, with few exceptions. From a health standpoint, they have the ability to cause a wide range of detrimental health effects. In humans, some of these compounds damage the kidneys, liver, and heart, others depress the central nervous system, and several are suspected to cause cancer. If human beings are subject to such health hazards, the logical question that follows is: What about their impact on delicate ecosystems?

ORGANIC COMPOUNDS

The molecules of organic compounds have one feature in common: one or more carbon atoms that covalently bond to other atoms, that is, pairs of electrons are shared between atoms. A carbon atom may share electrons with other nonmetallic atoms and also with other carbon atoms. As Figure 5.2 shows, methane, carbon tetrachloride,

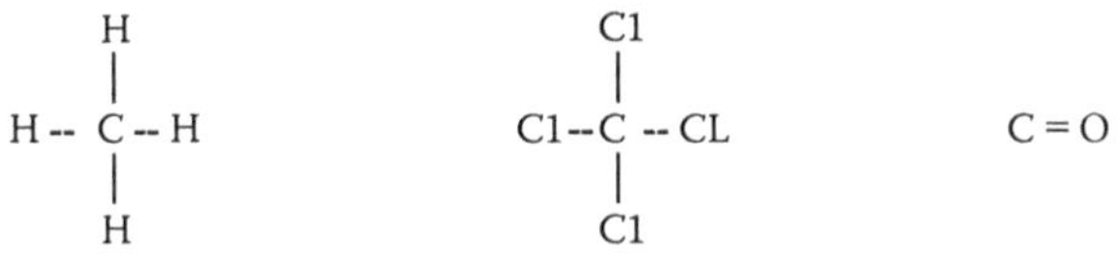

$$
\begin{array}{ccc}
\text{H} & \text{Cl} & \\
| & | & \\
\text{H -- C -- H} & \text{Cl -- C -- CL} & \text{C = O} \\
| & | & \\
\text{H} & \text{Cl} & \\
\end{array}
$$

Methane   **Carbon tetrachloride**   **Carbon monoxide**

FIGURE 5.2 Carbon atoms sharing their electrons with the electrons of other nonmetallic atoms, such as hydrogen, chlorine, and oxygen.

and carbon monoxide are compounds having moles in which the carbon atom is bonded to other nonmetallic atoms.

Did You Know?

Covalent bonds between carbon atoms in molecules or more complex organic compounds may be linked into chains, including branched chains, or into rings.

HYDROCARBONS

The simplest organic compounds are the *hydrocarbons*, compounds whose molecules are composed only of carbon and hydrogen atoms. All hydrocarbons are broadly divided into two groups: aliphatic and aromatic hydrocarbons.

Aliphatic hydrocarbons are those that can be characterized by the chain arrangements of their constituent carbon atoms. They are divided into the alkanes, alkenes, and alkynes.

The *alkanes* (also called paraffins or aliphatic hydrocarbons) are saturated hydrocarbons (hydrogen content is at maximum) with the general formula C_nH_{2n+2}. In systematic chemical nomenclature, alkane names end in the suffix -*ane*. They form the alkane series methane (CH_4), ethane (C_2H_6), propane (C_3H_8), butane (C_4H_{10}), etc. The lower members of the series are gases; the high-molecular-weight alkanes are waxy solids. Alkanes are present in natural gas and petroleum.

Alkenes (olefins) are unsaturated hydrocarbons (can take on hydrogen atoms to form saturated hydrocarbons) containing one or more double carbon-carbon bonds in their molecules. In systematic chemical nomenclature alkene names end in the suffix -*ene*. Alkenes with only one double bond form the alkene series starting with ethene (the gas that is liberated when food rots; ethylene), $CH_2:CH_2$; propene, $CH_3CH:CH_2$; etc.

Alkynes (acetylenes) are unsaturated hydrocarbons that contain one or more triple carbon-carbon bonds in their molecules. In systematic chemical nomenclature alkyne names end in the suffix -*yne*, such as acetylene, $H—C{\equiv}C—H$.

Aromatic hydrocarbons are those unsaturated organic compounds that contain a benzene ring in their molecules or that have chemical properties similar to those of benzene, which is a clear, colorless, water-insoluble liquid. Benzene, whose molecular

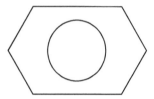

FIGURE 5.3 Molecular structure of benzene.

formula is C_6H_6, rapidly vaporizes at room temperature. The molecular structure of benzene is commonly represented by a hexagon with a circle inside, as shown in Figure 5.3.

ENVIRONMENTAL CHEMISTRY

Environmental chemistry is a blend of aquatic, atmospheric, soil chemistry, and the "chemistry" generated by human activities in those areas. The focus of this text, of course, deals with the environmental medium air. The environmental effects brought about by human influence cannot be totally ignored.

At this point, we understand that the chemistry that makes up the air medium—as well as the chemical reactions that take place to preserve and destroy this medium— is vital information in our study of environmental science. As we proceed through our treatment of air, we will be concerned with the environmental impact of human activities—mining, acid rain, erosion from poor cultivation practices, disposal of volatile hazardous wastes, photochemical reactions (smog), air pollutants such as particulate matter, the greenhouse effect, and ozone and water degradation problems related to organic, inorganic, and biological pollutants. All of these activities and problems have something to do with chemistry, and remediation or mitigation processes to solve them are also tied to chemistry.

Having said this, we turn to a discussion of atmospheric chemistry—the real core of air chemistry. To say that environmental science, environmental studies, and environmental engineering related to air science are built upon a strong foundation of chemistry is to mildly understate chemistry's real importance and relevance.

ATMOSPHERIC CHEMISTRY AND RELATED PHENOMENA

Note: In the brief discussion that follows describing the chemistry of earth's atmosphere, keep in mind that the atmosphere as it is at present (during the age of humans) is what is referred to. The atmosphere previous to this period was chemically quite different. Note also that atmospheric chemistry is a scientific discipline or entity that can stand on its own. The nuts and bolts of atmospheric chemistry are beyond the scope of this text. Here certain important atmospheric chemistry phenomena are highlighted.

You should recall that the full range of chemistry (including slow and fast reactions, dissolving crystals, and precipitation of solids) all occurs in the atmosphere, and that the atmosphere is as Graedel and Crutzen (1995) described it—a "flask without walls." We also pointed out that (excluding highly variable amounts of water vapor) more than 99% of the molecules constituting earth's atmosphere are nitrogen, oxygen, and chemically inert gases (noble gases such as argon, etc).

The chemistry (and thus the reactivity) of these natural gases (nitrogen, oxygen, carbon dioxide, argon, and others) is well known. The other reactive chemicals (anthropogenically produced) that are part of earth's atmosphere are also known, but there are still differing opinions on their exact total effect on our environment. For example, methane is by far the most abundant reactive compound in the atmosphere and currently is at a ground-level concentration (in the northern hemisphere) of about 1.7 ppmv. We know significant amounts of information about methane (its generation and fate when discharged) and its influence on the atmosphere; however, we are still conducting research to find out more.

Many different reactive molecules (other than methane) exist in the atmosphere. We may not be familiar with each of these reactants, but many of us certainly are familiar with their consequences: the ozone hole, greenhouse effect and global warming, smog, acid rain, the rising tide, and so on. It may surprise you to know, however, that the total amount of all these reactants in the atmosphere is seldom more than 10 ppmv anywhere in the world at any given time. The significance should be obvious: the atmospheric problems currently occurring on earth at this time are the result of less than one thousandth of 1% of all of the molecules in the atmosphere. This indicates that environmental damage (causing global atmospheric problems) can result from far less than the tremendous amounts of reactive substances we imagine are dangerous.

The quality of the air we breathe, the visibility and atmospheric esthetics, and our climate (all of which are dependent upon chemical phenomena that occur in the atmosphere) are important to our health and to our quality of life. Global atmospheric problems, such as the nature and level of air pollutants, concern the air science practitioner the most because they affect our health and quality of life.

Let us take a look at some of the important chemical species and their reactions within (primarily) the stratosphere of our atmosphere.

PHOTOCHEMICAL REACTION: SMOG PRODUCTION

A *photochemical reaction*, generally, is any chemical reaction in which light is produced or light initiates the reaction. Light can initiate reactions by exciting atoms or molecules and making them more reactive. The light energy becomes converted to chemical energy.

The photochemical reaction we are concerned with here is the action or absorption of electromagnetic solar radiation (light) by chemical species, which causes the reactions. The ability of electromagnetic radiation to cause photochemical reactions to occur is a function shown in the following relationship:

$$E = hv \qquad (5.6)$$

where
E = energy of a photon
v = frequency
h = Planck's constant, 6.62×10^{-27}

The major photochemical reaction we are concerned with is the one that produces photochemical smog. *Photochemical smog* is initiated by nitrogen dioxide, which absorbs the visible or ultraviolet energy of sunlight, forming nitric oxide to free atoms of oxygen (O), which then combine with molecular oxygen (O_2) to form *ozone* (O_3). In the presence of hydrocarbons (other than methane) and certain other organic compounds, a variety of chemical reactions take place. Some 80 separate reactions have been identified or postulated.

The photochemical reaction that produces the smog we are familiar with is dependent on two factors:

1. Smog concentration is linked to both the amount of sunlight and hydro-carbons present.
2. The amount is dependent on the initial concentration of nitrogen oxides.

In the production of photochemical smog, many different substances are formed in sequence, including acrolein, formaldehyde, and PAN (peroxyacetylnitrate). Photochemical smog (the characteristic haze of minute droplets) is a result of condensed low-volatility organic compounds. The organics irritate the eye, and also, together with ozone, can cause severe damage to leafy plants. Photochemical smog tends to be most intense in the early afternoon when sunlight intensity is greatest. It differs from traditional smog (Los Angeles–type smog), which is most intense in the early morning and is dispersed by solar radiation.

In addition to the photochemical reactions that produce smog, many other chemical reactions take place in earth's atmosphere, including ozone production, production of free radicals, chain reactions, oxidation processes, acid-base reactions, and many others.

REFERENCES AND RECOMMENDED READING

Graedel, T. E., and Crutzen, P. J. 1995. *Atmosphere, climate, and change.* New York: Scientific American.
Meyer, E. 1989. *Chemistry of hazardous materials*, 2nd ed. Englewood Cliffs, NJ: Prentice Hall.
Spellman, F. R., and Whiting, N. 2006. *Environmental science and technology: Concepts and applications.* 2nd ed. Rockville, MD: Government Institutes.

Part II

Atmospheric Science

6 The Atmosphere

This most excellent canopy, the air, look you, this brave o'erhanging firmament, this majestical roof fretted with golden fire ...

Shakespeare, *Hamlet*

INTRODUCTION

Several theories of cosmogony attempt to explain the origin of the universe. Without speculating on the validity of any one theory, the following is simply the author's view.

The time: 4,500 million years ago.

Before the universe there was time. Only time; otherwise, the vast void held only darkness—everywhere.

Overwhelming darkness.

Not dim, not murky, not shadowy, or not unlit. Simple nothingness—nothing but darkness, a shade of black so intense we cannot fathom or imagine it today. Light had no existence—this was black of blindness, of burial in the bowels of the earth, the blackness of no other choice.

With time—eons of time—darkness came to a sudden, smashing, shattering, annihilating, scintillating, cataclysmic end—and there was light, light everywhere. This new force replaced darkness and lit up the expanse without end, creating a brightness fed by billions of glowing round masses so powerful as to renounce and overcome the darkness that had come before.

With the light was heat energy that shone and warmed and transformed into mega-trillions of super-excited ions, molecules, and atoms—heat of unimaginable proportions, forming gases—gases we do not even know how to describe, how to quantify, let alone how to name. But gases they were—and they were everywhere.

With light, energy, heat, and gases present, the stage was set for the greatest show of all time, anywhere, ever: the formation of the universe.

Over time—time in stretches we cannot imagine, so vast we cannot contemplate them meaningfully—the heat, the light, the energy and gases all came together and grew, like an expanding balloon, into one solid glowing mass. But it continued to grow, with the pangs, sweating, and moans accompanying any birthing, until it had reached the point of no return—explosion level. And it did; it exploded with the biggest bang of all time (with the biggest bang hopefully of all time).

The Big Bang sent masses of hot gases in all directions—to the farthest reaches of anything, everything—into the vast, wide, measureless void. Clinging together as they rocketed, soared, and swirled, forming galaxies that gradually settled into their arcs through the void, constantly propelled away from the force of their origin, these masses began their eternal evolution.

Two masses concern us: the sun and earth.

Forces well beyond the power of the sun (beyond anything imaginable) stationed this massive gaseous orb approximately 93,000,000 miles from the dense molten core enveloped in cosmic gases and the dust of time that eventually became the insignificant mass we now call earth.

Distant from the sun, earth's mass began to cool, slowly; the progress was slower than we can imagine, but cool it did. While the dust and gases cooled, earth's inner core, mantle, and crust began to form—no more a quiet or calm evolution than the revolution that cast it into the void had been.

Downright violent was this transformation—the cooling surface only a facade for the internal machinations going on inside, outgassing from huge, deep destructive vents (we would call them volcanoes today) erupting continuously—never stopping, blasting away, delivering two main ingredients: magma and gas.

The magma worked to form the primitive features of earth's early crust. The gases worked to form earth's initial atmosphere—our point of interest: the atmosphere. Without atmosphere, what is there?

About 4 billion years before present, earth's early atmosphere was chemically reducing, consisting primarily of methane, ammonia, water vapor, and hydrogen—for life as we know it today, an inhospitable brew.

Earth's initial atmosphere was not a calm, quiet, quiescent environment; to the contrary, it was an environment best characterized as dynamic, ever changing, where bombardment after bombardment by intense, bond-breaking ultraviolet light, along with intense lightning and radiation from radionuclides, provided energy to bring about chemical reactions that resulted in the production of relatively complicated molecules, including amino acids and sugars (building blocks of life).

About 3.5 billion years before present, primitive life formed in two radically different theaters: on Earth and below the primordial seas near hydrothermal vents that spotted the wavering, water-covered floor.

Initially, on earth's unstable surface, these very primitive life-forms derived their energy from fermentation of organic matter formed by chemical and photochemical processes, then gained the ability to produce organic matter (CH_2O) by photosynthesis.

Thus, the stage was set for the massive biochemical transformation that resulted in the production of almost all the atmosphere's O_2.

The O_2 initially produced was quite toxic to primitive life-forms. However, much of this oxygen was converted to iron oxides by reaction with soluble iron. This process formed enormous deposits of iron oxides—the existence of which provides convincing evidence for the liberation of O_2 in the primitive atmosphere.

Eventually, enzyme systems developed that enabled organisms to mediate the reaction of waste-product oxygen with oxidizable organic matter in the sea. Later, the mode of waste gradient disposal was utilized by organisms to produce energy by respiration, which is now the mechanism by which nonphotosynthetic organisms obtain energy. In time, O_2 accumulated in the atmosphere. In addition to providing an abundant source of oxygen for respiration, the accumulated atmospheric oxygen formed an ozone (O_3) shield—the O_3 shield absorbs bond-rupturing ultraviolet radiation.

With the O_3 shield protecting tissue from destruction by high-energy ultraviolet radiation, the earth, although still hostile to life-forms we are familiar with, became a much more hospitable environment for life (self-replacing molecules), and life-forms were enabled to move from the sea (where they flourished next to the hydrothermal gas vents) to the land. And from that point on to the present, earth's atmosphere become more life-form friendly.

EARTH'S THIN SKIN

Shakespeare likened it to a majestic overhanging roof (constituting the transition between its surface and the vacuum of space); others have likened it to the skin of an apple. Both these descriptions of our atmosphere are fitting, as well as it being described as the earth's envelope, veil, or gaseous shroud. The atmosphere is more like the apple skin, however. This thin skin, or layer, contains the life-sustaining oxygen (21%) required by all humans and many other life-forms; the carbon dioxide (0.03%) so essential for plant growth; the nitrogen (78%) needed for chemical conversion to plant nutrients; the trace gases such as methane, argon, helium, krypton, neon, xenon, ozone, and hydrogen; and varying amounts of water vapor and airborne particulate matter. Life on earth is supported by this atmosphere, solar energy, and other plant's magnetic fields.

Gravity holds about half the weight of a fairly uniform mixture of these gases in the lower 18,000 feet of the atmosphere; approximately 98% of the material in the atmosphere is below 100,000 feet.

Atmospheric pressure varies from 1,000 millibars (mb) at sea level to 10 mb at 100,000 feet. From 100,000 to 200,000 feet the pressure drops from 9.9 mb to 0.1 mb, and so on.

The atmosphere is considered to have a thickness of 40–50 miles; however, here we are primarily concerned with the troposphere, the part of the earth's atmosphere that extends from the surface to a height of about 27,000 feet above the poles, about 36,000 feet in mid-latitudes, and about 53,000 feet over the equator. Above the troposphere is the stratosphere, a region that increases in temperature with altitude (the warming is caused by absorption of the sun's radiation by ozone) until it reaches its upper limit of 260,000 feet.

THE TROPOSPHERE

Extending above earth approximately 27,000 feet, the troposphere is the focus of this text because people, plants, animals, and insects live here and depend on this thin layer of gases. Moreover, all of the earth's weather takes place within the troposphere. The troposphere begins at ground level and extends 7.5 miles up into the sky, where it meets with the second layer called the stratosphere.

Did You Know?

It was pointed out earlier that the gases that are so important to life on earth are primarily contained in the troposphere. Also note that another important substance is contained in the troposphere: water vapor. Along with being the most remarkable of the trace gases contained in the troposphere, water vapor is also the most variable. Unlike the other trace gases in the atmosphere, water vapor alone exists in gas, solid, and liquid forms. It also functions to add and remove heat from the air when it changes from one form to another.

Water vapor (in conjunction with airborne particles, obviously) is essential for the stability of earth's ecosystem. This water vapor–particle combination interacts with the global circulation of the atmosphere and produces the world's weather, including clouds and precipitation.

THE STRATOSPHERE

The stratosphere begins at the 7.5-mile point and reaches 21.1 miles into the sky. In the rarified air of the stratosphere, the significant gas is ozone (life-protecting ozone—not to be confused with pollutant ozone), which is produced by the intense ultraviolet radiation from the sun. In quantity, the total amount of ozone in the atmosphere is so small that if it were compressed to a liquid layer over the globe at sea level, it would have a thickness of less than 3/16 inch.

Ozone contained in the stratosphere can also impact (add to) ozone in the troposphere. Normally, the troposphere contains about 20 parts per billion parts of ozone. On occasion, however, via the jet stream, this concentration can increase to five to ten times higher than average.

In our discussion of the earth's atmosphere in this book the focus is on the troposphere and stratosphere because these two layers directly impact life as we know it and are or can be heavily influenced by pollution and its effects.

Did You Know?

The troposphere, stratosphere, mesosphere, and thermosphere act together as a giant safety blanket. They keep the temperature on the earth's surface from dipping to extreme icy cold that would freeze everything solid, or from soaring to blazing heat that would burn up all life.

A JEKYLL AND HYDE VIEW OF THE ATMOSPHERE

When noncity dwellers look up into that great natural canopy above our heads, they see many features provided by our world's atmosphere that we know and enjoy: the blueness and clarity of the sky, the color of a rainbow, the spattering of stars reaching every corner of blackness, the magical colors of a sunset. The air they breathe carries the smell of ocean air, the refreshing breath of clean air after a thunderstorm, and the beauty contained in a snowflake.

But the atmosphere sometimes presents another face—Mr. Hyde's face. The terrible destructiveness of a hurricane, tornado, monsoon, typhoon, or hailstorm, the wearying monotony of winds carrying dusts and rampaging windstorms carrying fire up a hillside—these are some of the terrifying aspects of the other face of the disturbed atmosphere.

The atmosphere can also present a Hyde-like face whenever humans are allowed to pour their filth (pollution) into it: a view afforded from patches here and there that are not blocked by their buildings, their pollution that rising from their enterprises can mask the stars and make the visible sky a dirty yellow-brown or at best a sickly pale blue.

Fortunately, earth's atmosphere is self-healing. Air cleaning is provided by clouds, and the global circulation system constantly purges the air of pollutants. Only when air pollutants overload nature's way of rejuvenating its systems to their natural state are we faced with repercussions that can be serious, even life threatening.

ATMOSPHERIC PARTICULATE MATTER

Along with gases and water vapor earth's atmosphere is literally a boundless arena for particulate matter of many sizes and types. Atmospheric particulates vary in size from 0.0001 to 10,000 microns. Particulate size and shape have a direct bearing on visibility. For example, a spherical particle in the 0.6-micron range can effectively scatter light in all directions, reducing visibility.

The types of airborne particulates in the atmosphere vary widely, with the largest sizes derived from volcanoes, tornados, waterspouts, burning embers from forest fires, and seed parachutes, spider webs, pollen, soil particles, and living microbes.

The smaller particles (the ones that scatter light) include fragments of rock, salt and spray, smoke, and particles from forested areas. The largest portion of airborne particulates is invisible. They are formed by the condensation of vapors, chemical reactions, photochemical effects produced by ultraviolet radiation, and ionizing forces that come from radioactivity, cosmic rays, and thunderstorms.

Airborne particulate matter is produced either by mechanical weathering, breakage, and solution or by the vapor-to-condensation-to-crystallization process (typical of particulates from a furnace of a coal-burning power plant).

As you might guess, anything that goes up must eventually come down. This is typical of airborne particulates also. Fallout of particulate matter depends, obviously, mostly on their size, and less obviously, on their shape, density, weight, airflow, and injection altitude. The residence time of particulate matter also is dependent on the atmosphere's cleanup mechanisms (formation of clouds and precipitation) that work to remove them from their suspended state.

Some large particulates may only be airborne for a matter of seconds or minutes, with intermediate sizes able to stay afloat for hours or days. The finer particulates may stay airborne for a much longer duration: for days, weeks, months, and even years.

Particles play an important role in atmospheric phenomena. For example, particulates provide the nuclei upon which ice particles are formed, cloud condensation forms, and condensation takes place. Obviously, the most important role airborne particulates play is in cloud formation. Simply put, without clouds life as we know it would be much more difficult, and cloud bursts that eventually erupted would cause such devastation that it is hard to imagine or contemplate.

The situation just described could also result whenever massive forest fires and volcanic action take place. These events would release a superabundance of cloud condensation nuclei, which would overseed the clouds, causing massive precipitation to occur. If natural phenomena such as forest fires and volcanic eruptions can overseed clouds and cause massive precipitation, then what effect would result from man-made pollutants entering the atmosphere at unprecedented levels? This question and other pollution-related questions will be answered in subsequent chapters.

REFERENCES AND RECOMMENDED READING

EPA. 2007. Air pollution control orientation course: Air pollution. www.epa.gov/air/oaqps/ eog/course422/ap.1.html (accessed January 5, 2008).

Spellman, F. R., and Whiting, N. 2006. *Environmental science and technology: Concepts and applications*. 2nd ed. Rockville, MD: Government Institutes.

7 Moisture in the Atmosphere

Hath the rain a father? or who hath begotten the drops of dew? Out of whose
womb came the ice? and the hoary frost of heaven, who hath gendered it? …
Can't thou lift up thy voice to the clouds, that abundance of water may cover
thee? (Job 38:28–29, 34)

I wondered lonely as a cloud
That floats on high o'er vales and hills,
When all at once I saw a crowd,
A host, of golden daffodils;
Beside the lake, beneath the trees,
Fluttering and dancing in the breeze.

William Wordsworth, 1804

INTRODUCTION

On a hot day when clouds build up, signifying that a storm is imminent, we do not
always appreciate what is happening.

What is happening?

This cloud buildup actually signals that one of the most vital processes in the
atmosphere is occurring: the condensation of water as it is raised to higher levels
and cooled within strong updrafts of air created by either convection currents, turbulence, or physical obstacles like mountains. The water originated from the surface—evaporated from the seas, from the soil, or transpired by vegetation. Once
within the atmosphere, however, a variety of events combine to convert the water
vapor (produced by evaporation) to water droplets. The air must rise and cool to
its dew point, of course. At dew point, water condenses around minute airborne
particulate matter to make tiny cloud droplets, forming clouds—clouds from which
precipitation occurs.

Whether created by the sun heating up a hillside, jet aircraft exhausts, or factory
chimneys, there are actually only ten major cloud types. The deliverers of countless
millions of tons of moisture from the earth's atmosphere, they form even from the
driest desert air containing as little as 0.1% water vapor. They not only provide a visible sign of motion, but also indicate change in the atmosphere portending weather
conditions that may be expected up to 48 hours ahead. In this chapter we take a brief
look at the nature and consequences of these cloud-forming processes.

CLOUD FORMATION

The atmosphere is a highly complex system, and the effects of the changes in any single property tend to be transmitted to many other properties. The most profound effect on the atmosphere is the result of alternate heating and cooling of the air, which causes adjustments in relative humidity and buoyancy; they cause condensation, evaporation, and cloud formation.

The temperature structure of the atmosphere (along with other forces that propel the moist air upward) is the main force behind the form and size of clouds. Exactly how does temperature affect atmospheric conditions? For one thing, temperature (that is, heating and cooling of the surface atmosphere) causes vertical air movements. Let us take a look at what happens when air is heated.

Let us start with a simple parcel of air in contact with the ground. As the ground is heated, the air in contact with it will warm also. This warm air increases in temperature and expands. Remember, gases expand on heating much more than liquids or solids, so this expansion is quite marked. In addition, as the air expands, its density falls (meaning that the same mass of air now occupies a larger volume). You've heard that warm air rises? Because of its lessened density, this parcel of air is now lighter than the surrounding air and tends to rise. Conversely, if the air cools, the opposite occurs—it contracts, its density increases, and it sinks. Actually, alternate heating and cooling are intimately linked with the process of evaporation, condensation, and precipitation.

But how does a cloud actually form? Let us look at another example. On a sunny day, some patches of ground warm up more quickly than others because of differences in topography (soil and vegetation, etc.). As the surface temperature increases, heat passes to the overlying air. Later, by mid-morning, a bulbous mass of warm, moisture-laden air rises from the ground. This mass of air cools as it meets lower atmospheric pressure at higher altitudes. If cooled to its dew point temperature, condensation follows and a small cloud forms. This cloud breaks free from the heated patch of ground and drifts with the wind. If it passes over other rising air masses, it may grow in height. The cloud may encounter a mountain and be forced higher still into the air. Condensation continues as the cloud cools, and if the droplets it holds become too heavy, they fall as rain.

Did You Know?

Clouds play an important role in boundary layer meteorology and air quality. Convective clouds transport pollutants vertically, allowing an exchange of air between the boundary layer and the free troposphere. Cloud droplets formed by heterogeneous nucleation on aerosols grow into rain droplets through condensation, collision, and coalescence. Clouds and precipitation scavenge pollutants from the air. Once inside the cloud or rainwater, some compounds dissociate into ions or react with one another through aqueous chemistry. Another important role for clouds is the removal of pollutants trapped in rainwater and its deposition onto the ground (EPA/600/R-99/030).

MAJOR CLOUD TYPES

Earlier, it was mentioned that there are ten major cloud types. These include:

Genus: *Stratiform*
  Species:
  Cirrus
  Cirrostratus
  Cirrocumulus
  Altostratus
  Altocumulus
  Stratus
  Stratocumulus
  Nimbostratus
Genus: *Cumuliform*
  Species:
  Cumulus
  Cumulonimbus

From the list above it is apparent that the cloud groups are classified into a system that uses Latin words to describe the appearance of clouds as seen by an observer on the ground. Table 7.1 summarizes the four principal components of this classification system (Ahrens, 1994).

Further classification identifies clouds by height of cloud base. For example, cloud names containing the prefix *cir-*, as in cirrus clouds, are located at high levels, while cloud names with the prefix *alto-*, as in altostratus, are found at middle levels. This module introduces several cloud groups. The first three groups are identified based upon their height above the ground. The fourth group consists of vertically developed clouds, while the final group consists of a collection of miscellaneous cloud types.

Let us take a closer look at each of these cloud types.

A *stratus* cloud is a featureless, gray, low-level cloud. Its base may obscure hilltops or occasionally extend right down to the ground, and because of its low altitude, it appears to move very rapidly on breezy days. Stratus can produce drizzle

TABLE 7.1
Summary of Components of Cloud Classification System

Latin Root	Translation	Example
Cumulus	Heaped/puffy	Fair weather cumulus
Stratus	Layered	Altostratus
Cirrus	Curl of hair/wispy	Cirrus
Nimbus	Rain	Cumulonimbus

or snow, particularly over hills, and may occur in huge sheets covering several thousand miles.

Cumulus clouds also seem to scurry across the sky, reflecting their low altitude. These small, dense, white, fluffy, flat-based clouds are typically short lived, lasting no more than 10–15 minutes before dispersing. They are typically formed on sunny days, when localized convection currents are set up. These currents can form over factories or even brush fires, which may produce their own clouds.

Cumulus may expand into low-lying horizontally layered, massive *stratocumulus*, or into extremely dense, vertically developed, giant *cumulonimbus*, with a relatively hazy outline and a glaciated top that are up to 7 miles in diameter. These clouds typically form on summer afternoons; their high, flattened tops contain ice, which may fall to the ground in the form of heavy showers of rain or hail.

Rising to middle altitudes, the bluish gray layered *altostratus* and rounded, fleecy, whitish gray *altocumulus* appear to move slowly because of their greater distance from the observer.

Cirrus (meaning tuft of hair) clouds are made up of white narrow bands of thin, fleecy parts and are relatively common over Northern Europe, and generally ride the jet stream rapidly across the sky.

Cirrocumulus are high-altitude clouds composed of a series of small, regularly arranged cloudlets in the form of ripples or grains; they are often present with cirrus clouds in small amounts. *Cirrostratus* are high-altitude, thin, hazy clouds, usually covering the sky and giving a halo effect surrounding the sun or moon.

Did You Know?

Clouds whose names incorporate the word *nimbus* or the prefix *nimbo-* are clouds from which precipitation is falling.

MOISTURE IN THE ATMOSPHERE

Let us summarize the information related to how moisture accumulates in and precipitates from the atmosphere. The process of evaporation (converting moisture into vapor) supplies moisture into the lower atmosphere. The prevailing winds then circulate the moisture and mix it with drier air elsewhere.

Water vapor is only the first stage of the precipitation cycle; the vapor must be converted into liquid form. This is usually achieved by cooling, either rapidly, as in convection, or slowly, as in cyclonic storms. Mountains also cause uplift, but the rate will depend upon their height and shape and the direction of the wind.

To actually produce precipitation, the cloud droplets must become large enough to reach the ground without evaporating. The cloud must possess the right physical properties to enable the droplets to grow.

If the cloud lasts long enough for growth to take place, then precipitation will usually occur. Precipitation results from a delicate balance of counteracting forces, some leading to droplet growth and others to droplet destruction.

Did You Know?

Contrails are clouds formed around the small particles (aerosols) that are in air-craft exhaust. When these persist after the passage of the plane, they are indeed clouds, and are of great interest to researchers. Under the right conditions, clouds initiated by passing aircraft can spread with time to cover the whole sky.

REFERENCES AND RECOMMENDED READING

Ahrens, D. 1994. *Meteorology today: An introduction to weather, climate and the environment.* 5th ed. St. Paul, MN: West Publishing Company.

EPA. 2007. Air pollution control—Atmosphere. www.epa.gov/air/oaqpseog course422/apl. html (accessed December 28, 2007).

NASA. 2008. *Observing cloud type.* Washington, DC: Author.

NOAA. 2007. Cloud types. www.gfdl.NOAA.gov/~01/weather/clouds.html (accessed December 29, 2007).

Spellman, F. R., and Whiting, N. 2006. *Environmental science and technology: Concepts and applications.* 2nd ed. Rockville, MD: Government Institutes.

8 Precipitation and Evapotranspiration

Because it determines the intensity and distribution of many of the processes operating within the system, precipitation is one of the most important regulators of the hydrological cycle. The rate of evapotranspiration is closely related to precipitation, and thus is also an integral part of the hydrological cycle.

The Rainy Day

> The day is cold, and dark, and dreary;
> It rains, and the wind is never weary;
> The vine still clings to the moldering wall,
> But at every gust the dead leaves fall,
> And the day is dark and dreary.
>
> My life is cold, and dark, and dreary;
> It rains, and the wind is never weary;
> My thoughts still cling to the moldering Past,
> But the hopes of youth fall thick in the blast
> And the days are dark and dreary.
>
> Be still, sad heart! And cease repining;
> Behind the clouds is the sun still shining;
> Thy fate is the common fate of all,
> Into each life some rain must fall,
> Some days must be dark and dreary.

Henry Wadsworth Longfellow

INTRODUCTION

The principal actions brought on by weather systems that affect land and sea and the humans, animals, and vegetation thereon are winds and precipitation. The latter comes in a variety of forms, as discussed below. Most weather of consequence to people occurs in storms. These may be local in origin but more commonly are carried to locations in wide areas along pathways followed by active air masses consisting of highs and lows. The key ingredient in storms is water, as either a liquid or a vapor. The vapor acts like a gas and thus contributes to the total pressure of the atmosphere, making up a small but vital fraction of the total (NASA, 2008).

Precipitation is found in a variety of forms. Which form actually reaches the ground depends upon many factors, for example, atmospheric moisture content, surface temperature, intensity of updrafts, and method and rate of cooling.

Water vapor in the air will vary in amount depending on sources, quantities, processes involved, and air temperature. Heat, mainly as solar irradiation but with some contributed by the earth and human activity and some from change-of-state processes, will cause some water molecules, either in water bodies (oceans, lakes, rivers) or in soils, to be excited thermally and escape from their sources. This is called evaporation. When water is released from trees and other vegetation, the process is known as evapotranspiration. The evaporated water, or moisture, that enters the air is responsible for a state called humidity. Absolute humidity is the weight of water vapor contained in a given volume of air. The mixing ratio refers to the mass of the water vapor within a given mass of dry air. At any particular temperature, the maximum amount of water vapor that can be contained is limited to some amount; when that amount is reached, the air is said to be saturated for that temperature. If less than the maximum amount is present, then the property of air that indicates this is its relative humidity (RH), defined as the actual water vapor amount compared to the saturation amount at the given temperature; this is usually expressed as a percentage. RH also indicates how much moisture the air can hold above its stated level, which, after attaining, could lead to rain.

When a parcel of air attains or exceeds RH = 100%, condensation will occur and water in some state will begin to organize as some type of precipitation. One familiar form is dew, which occurs when the saturation temperature or some quantity of moisture reaches a temperature at the surface at which condensations sets in, leaving the moisture to coat the ground (especially obvious on lawns).

The term *dew point* has a more general use, being that temperature at which an air parcel must be cooled to become saturated. Dew frequently forms when the current air mass contains excessive moisture after a period of rain but the air is now clear; the dew precipitates out to coat the surface (noticeable on vegetation). Ground fog is a variant in which lower temperatures bring on condensation within the near surface air as well as the ground.

The other types of precipitation are listed in Table 8.1, along with descriptive characteristics related to each type.

Evaporation and transpiration are complex processes that return moisture to the atmosphere. The rate of evapotranspiration depends largely on two factors: (1) how saturated (moist) the ground is and (2) the capacity of the atmosphere to absorb the moisture. In this chapter we discuss the factors responsible for both precipitation and evapotranspiration.

PRECIPITATION

In Chapter 7 it was stated that if all the essentials were present, precipitation occurs when the dew point is reached. However, Chapter 7 also pointed out that it is quite possible for an air mass or cloud containing water vapor to be cooled below the dew point without precipitation occurring. In this state the air mass is said to be *supercooled*.

TABLE 8.1
Types of Precipitation

Type	Approximate Size	State of Water	Description
Mist	0.005–0.05 mm	Liquid	Droplets large enough to be felt on face when air is moving 1 meter/second; associated with stratus clouds
Drizzle	Less than 0.5 mm	Liquid	Small uniform drops that fall from stratus clouds, generally for several hours
Rain	0.5–5 mm	Liquid	Generally produced by nimbostratus or cumulonimbus clouds; when heavy, size can be highly variable from one place to another
Sleet	0.5–5 mm	Solid	Small, spherical to lumpy ice particles that form when raindrops freeze while falling through a layer of subfreezing air; because the ice particles are small, any damage is generally minor; sleet can make travel hazardous
Glaze	Layers 1 mm–2 cm thick	Solid	Produced when supercooled raindrops freeze on contact with solid objects; glaze can form a thick covering of ice having sufficient weight to seriously damage trees and power lines
Rime	Variable accumulation	Solid	Deposits usually consisting of ice feathers that point into the wind; these delicate frostlike accumulations form as supercooled cloud or fog droplets encounter objects and freeze on contact
Snow	1 mm–2 cm	Solid	Crystalline nature allows it to assume many shapes, including six-sided crystals, plates, and needles; produced in supercooled clouds, where water vapor is deposited as crystals that remain frozen during their descent
Hail	5 mm to cm or larger	Solid	Precipitation in the form of hard, rounded pellets or irregular lumps of ice; produced in large convective, cumulonimbus clouds, where frozen ice particles and supercooled water coexist
Graupel	2–5 mm	Solid	Sometimes called "soft hail"; forms on rime and collects on snow crystals to produce irregular masses of "soft" ice; because these particles are softer than hailstones, they normally flatten out upon impact

Source: NASA (2008).

How, then, are droplets of water formed? Water droplets form around micro-scopic foreign particles already present in the air. These particles on which the drop-lets form are called *hygroscopic nuclei*. They are present in the air primarily in the form of dust, salt from seawater evaporation, and from combustion residue. These foreign particles initiate the formation of droplets that eventually fall as precipitation. To have precipitation, larger droplets or drops must form. This may be brought about by two processes: (1) coalescence (collision) or (2) the Bergeron process.

COALESCENCE

Simply put, *coalescence* is the fusing together of smaller droplets into larger ones. The variation in the size of the droplets has a direct bearing on the efficiency of this process. Raindrops come in different sizes and can reach diameters up to 7 millimeters. Having larger droplets greatly enhances the coalescence process.

But what actually goes on inside a cloud to cause rain to fall? To answer this question, we must take a look inside a cloud to see exactly what processes occur to make rain—rain that actually falls as rain. Rainmaking is based on the essentials of the Bergeron process.

BERGERON PROCESS

Named after the Swedish meteorologist who suggested it, the *Bergeron process* is probably the more important process for the initiation of precipitation. To gain an understanding on how the Bergeron process works, let us look at what actually goes on inside a cloud to cause rain.

Within a cloud made up entirely of water droplets there will be a variety of droplet sizes. The air will be rising within the cloud anywhere from 10–20 centimeters per second (depending on type of cloud). As the air rises, the drops become larger through collision and coalescence; many will reach drizzle size. Then the updraft intensifies up to 50 centimeters per second (and more), which reduces the downward movement of the drops, allowing them more time to become even larger. When the cloud becomes approximately 1 kilometer deep, small raindrops of 700 μm diameter are formed.

The droplets, because of their small size, do not freeze immediately, even when the temperatures fall below 0°C. Instead, the droplets remain unfrozen in a super-cooled state. However, when the temperature drops as low as –10°C, ice crystals may start to develop among the water droplets. This mixture of water and ice would not be particularly important but for a peculiar characteristic or property of water. Therefore, at –10°C, air saturated with respect to liquid water is supersaturated rela-tive to ice by 10%, and at –20°C by 21%. Thus, ice crystals in the cloud tend to grow and become heavier at the expense of the water droplets.

Eventually, the ice crystals sink to the lower levels of the cloud where temper-atures are only just below freezing. When this occurs, they tend to combine (the supercooled droplets of water act as an adhesive) and form snowflakes. When the snowflakes melt, the resulting water drops may grow further by collision with cloud droplets before they reach the ground as rain. The actual rate at which water vapor is converted to raindrops depends on three main factors: (1) the rate of ice crystal growth, (2) supercooled vapor, and (3) the strength of the updrafts (mixing) in the cloud.

Types of Precipitation

We stated that in order for precipitation to occur, water vapor must condense, which occurs when water vapor ascends and cools. Three mechanisms by which air rises allowing for precipitation to occur are convectional, orographic, and frontal.

Convectional Precipitation

Convectional precipitation is the spontaneous rising of moist air due to instability. This type of precipitation is usually associated with thunderstorms and occurs in the summer because localized heating is required to initiate the convection cycle. We have discussed that upward-growing clouds are associated with convection. Since the updrafts (commonly called a thermal) are usually strong, cooling of the air is rapid and lots of water can be condensed quickly, usually confined to a local area, and a sudden summer downpour may occur as a result.

Convectional thunderstorm clouds are also described as supercells. Convective thunderstorms are the most common type of atmospheric instability that produces lightning followed by thunder. Lightning is one of the most spectacular phenomena witnessed in storms.

Did You Know?

A lightning bolt can attain an electric potential up to 30 million volts and current as much as 10,000 amps. It can cause air temperatures to reach 10,000 °C. But a bolt's duration is extremely short (fractions of a second). Although a bolt can kill people it hits, most can survive.

Orographic Precipitation

Orographic precipitation is a straightforward process, characteristic of mountainous regions; almost all mountain areas are wetter than the surrounding lowlands. This type of precipitation arises when air is forced to rise over a mountain or mountain range. The wind, blowing along the surface of the earth, ascends along topographic variations. Where air meets this extensive barrier, it is forced to rise. This ascending wind usually gives rise to cooling and encourages condensation, and thus orographic precipitation, on the windward side of the mountain range.

Frontal Precipitation

Frontal precipitation results when two different fronts (or the boundary between two air masses characterized by varying degrees of precipitation), at different temperatures, meet. The warm air mass (since it is lighter) moves up and over the colder

---

TABLE 8.2
Water Balance in the United States (in bgd)

Precipitation	4,200
Evaporation and transpiration	3,000
Runoff	1,250
Withdrawal	310
Irrigation	142
Industry (utility cooling water)	142
Municipal	26
Consumed (irrigation loss)	90
Returned to streams	220

Source: National Academy of Sciences (1962).
Note: bgd, billion gallons per day.

---

air mass. The cooling is usually less rapid than in the vertical convection process because the warm air mass moves up at an angle, more of a horizontal motion.

EVAPOTRANSPIRATION

Another important part or process of the hydrological cycle (though it is often neglected because it can rarely be seen) is *evapotranspiration*. More complex than precipitation, evaporation and transpiration created a land-atmosphere interface process whereby a major flow of moisture is transferred from the ground level to the atmosphere. It returns moisture to the air, replenishing that lost by precipitation, and it also takes part in the global transfer of energy. The rate of evapotranspiration depends largely on two factors: (1) how moist the ground is and (2) the capacity of the atmosphere to absorb the moisture. Therefore, the greatest rates are over the tropical oceans, where moisture is always available and the long hours of sunshine and steady trade winds evaporate vast quantities of water.

Just how much moisture is returned to the atmosphere via transpiration? In answering this question, Table 8.2 makes clear, for example, that in the United States alone, about two-thirds of the average rainfall over the U.S. mainland is returned via evaporation and transpiration.

EVAPORATION

Evaporation is the process by which a liquid is converted into a gaseous state. Evaporation takes place (except when air reaches saturation at 100% humidity) almost on a continuous basis. It involves the movement of individual water molecules from the surface of earth into the atmosphere, a process occurring whenever a vapor pressure gradient exists from the surface to the air (i.e., whenever the humidity of the atmosphere is less than that of the ground). Evaporation also requires energy (derived from the sun or from sensible heat from the atmosphere or ground)—2.48×10^6 joules to evaporate each kilogram of water at 10°C.

TRANSPIRATION

A related process, *transpiration* is the loss of water from a plant by evaporation. Most water is lost from the leaves through pores known as stomata, whose primary function is to allow gas exchange between the plant's internal tissues and the atmosphere. Transpiration from the leaf surfaces causes a continuous upward flow of water from the roots via the xylem, which is known as the transpiration stream.

Transpiration occurs mainly by day, when the stomata open up under the influence of sunlight. Acting as evaporators, they expose the pure moisture (the plant's equivalent of perspiration) in the leaves to the atmosphere. If the vapor pressure of the air is less than that in the leaf cells, the water is transpired.

As you might guess, because of transpiration, far more water passes through a plant than is needed for growth. In fact, only about 1% or so is actually used in plant growth. Nevertheless, the excess movement of moisture through the plant is important to the plant because the water acts as a solvent, transporting vital nutrients from the soil into the roots and carrying them through cells of the plant. Obviously, without this vital process plants would die.

EVAPOTRANSPIRATION: THE PROCESS

Although evapotranspiration plays a vital role in cycling water over earth's land masses, it is seldom appreciated. In the first place, distinguishing between evaporation and transpiration is often difficult. Both processes tend to be operating together, so the two are normally combined to give the composite term *evapotranspiration*.

Governed primarily by atmospheric conditions, energy is needed to power the process. Wind also plays an important role, which acts to mix the water molecules with the air and transport them away from the surface. The primary limiting factor in the process is lack of moisture at the surface (soil is dry). Evaporation can continue only so long as there is a vapor pressure gradient between the ground and the air.

REFERENCES AND RECOMMENDED READING

NASA. 2008. *Hydrologic Cycle*. Accessed Jan. 8, 2008 @http://www.NASA.gov/audience/forstudents/5-8/features/observations.

National Academy of Sciences. 1962. *Water balance in the U.S.* National Research Council Publication 100-B.

Shipman, J. T., Adams, J. L., and Wilson, J. D. 1987. *An introduction to physical science.* Lexington, MA: D.C. Heath and Company.

Spellman, F. R. 2007. *The science of water.* 2nd ed. Boca Raton, FL: CRC Press.

USGS. 2008. The water cycle: Evapotranspiration. http://ga.water.usgs.gov/edu/watercycle.html (accessed January 7, 2008).

9 The Atmosphere in Motion

There are scientists and engineers in the real world who will tell us that perpetual motion in or for any machine is a pipedream—it is wishful thinking and impossible. Have you ever pondered the most dynamic perpetual motion machine of them all—earth's atmosphere?

INTRODUCTION

Have you ever wondered why the earth's atmosphere is in perpetual motion? Probably not, but it is. It must be in a state of perpetual motion because it constantly strives to eliminate the constant differences in temperature and pressure between different parts of the globe. How are these differences eliminated or compensated for? By its motion, which produces winds and storms. In this chapter the horizontal movements that transfer air around the globe are considered.

GLOBAL AIR MOVEMENT

Basically, winds are the movement of the earth's atmosphere, which by its weight exerts a pressure on the earth that we can measure using a barometer. Winds are often confused with air currents, but they are different. Wind is the horizontal movement of air or motion along the earth's surface. Air currents, on the other hand, are vertical air motions collectively referred to as updrafts and downdrafts.

Throughout history, man has been both fascinated by and frustrated by winds. Man has written about winds almost from the time of the first written word. For example, Herodotus (and later Homer and many others) wrote about winds in his *The Histories*. Wind has had such an impact upon human existence that humans have given names to describe particular winds, specific to a particular geographical area. Table 9.1 lists some of these winds, their colorful names, and the region where they occur. Some of these names are more than just colorful—the winds are actually colored. For example, the *harmattan* blows across the Sahara filled with red dust; mariners called this red wind the sea of darkness.

EARTH'S ATMOSPHERE IN MOTION

To state that earth's atmosphere is constantly in motion is to state the obvious. Anyone observing the constant weather changes around him or her is well aware of this phenomenon. Although obvious, the importance of the dynamic state of our atmosphere is much less obvious.

As mentioned, the constant motion of earth's atmosphere (air movement) consists of both horizontal (wind) and vertical (air currents) dimensions. The atmosphere's

---

TABLE 9.1
Assorted Winds of the World

Wind Name	Location
Aajej	Morocco
Alm	Yugoslavia
Biz roz	Afghanistan
Haboob	Sudan
Imbat	North Africa
Datoo	Gibraltar
Nafhat	Arabia
Besharbar	Caucasus
Samiel	Turkey
Tsumuji	Japan
Brickfielder	Australia
Chinook	America
Williwaw	Alaska

---

motion is the result of thermal energy produced from the heating of the earth's surface and the air molecules above. Because of differential heating of the earth's surface, energy flows from the equator pole-ward.

Even though air movement plays the critical role in transporting the energy of the lower atmosphere, bringing the warming influences of spring and summer and the cold chill of winter, the effects of air movements on our environment are often overlooked, even though wind and air currents are fundamental to how nature functions. All life on earth has evolved with mechanisms dependent on air movement: pollen is carried by winds for plant reproduction; animals sniff the wind for essential information; wind power was the motivating force that began the earliest stages of the industrial revolution. Now we see the effects of winds in other ways, too: wind causes weathering (erosion) of the earth's surface; wind influences ocean currents; air pollutants and contaminants such as radioactive particles transported by the wind impact our environment.

CAUSES OF AIR MOTION

In all dynamic situations, forces are necessary to produce motion and changes in motion—winds and air currents. The air (made up of various gases) of the atmosphere is subject to two primary forces: (1) gravity and (2) pressure differences from temperature variations.

Gravity (gravitational forces) holds the atmosphere close to the earth's surface. Newton's law of universal gravitation states that every body in the universe attracts another body with a force equal to:

$$F = G\frac{m_1 m_2}{R^2} \tag{9.1}$$

where

F = force

m_1 and m_2 = the masses of the two bodies

G = universal constant of 6.67×10^{-11} N × m²/kg²

R = distance between the two bodies

Important Point: The force of gravity decreases as an inverse square of the distance between them.

Thermal conditions affect density, which in turn cause gravity to affect vertical air motion and planetary air circulation. This affects how air pollution is naturally removed from the atmosphere.

Although forces in other directions often overrule gravitational force, the ever-present force of gravity is vertically downward and acts on each gas molecule, accounting for the greater density of air near the earth.

Atmospheric air is a mixture of gases, so the gas laws and other physical principles govern its behavior. The pressure of a gas is directly proportional to its temperature. Pressure is force per unit area (P = F/A), so a temperature variation in air generally gives rise to a difference in pressure of force. This difference in pressure resulting from temperature differences in the atmosphere creates air movement—on both large and local scales. This pressure difference corresponds to an unbalanced force, and when a pressure difference occurs, the air moves from a high- to a low-pressure region.

In other words, horizontal air movements (called advective winds) result from temperature gradients, which give rise to density gradients, and subsequently pressure gradients. The force associated with these pressure variations (pressure gradient force) is directed at right angles to (perpendicular to) lines of equal pressure (called isobars) and is directed from high to low pressure.

Look at Figure 9.1. The pressures over a region are mapped by taking barometric readings at different locations. Lines drawn through the points (locations) of equal pressure are called isobars. All points on an isobar are of equal pressure, which means no air movement along the isobar. The wind direction is at right angles to the isobar in the direction of the lower pressure. In Figure 9.1, notice that air moves down a pressure gradient toward a lower isobar like a ball rolls down a hill. If the isobars are close together, the pressure gradient force is large, and such areas are characterized by high wind speeds. If isobars are widely spaced (see Figure 9.1), the winds are light because the pressure gradient is small.

Did You Know?

Air pressure at any location, whether it is on the earth's surface or up in the atmosphere, depends on the weight of the air above. Imagine a column of air. At sea level, a column of air extending hundreds of kilometers above sea level exerts a pressure of 1,013 millibars (mb), or 1.013 L. But, if you travel up the column to an altitude of 4.4 kilometers (18,000 feet), the air pressure would be roughly half, or approximately 506 millibars.

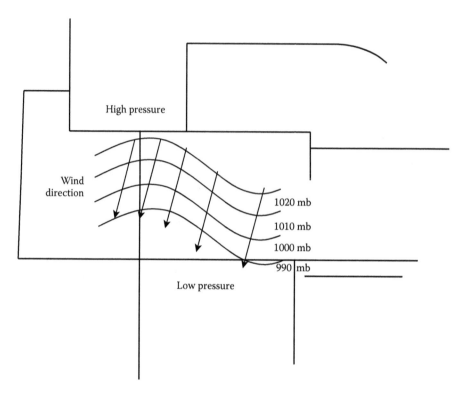

FIGURE 9.1 Isobars drawn through locations having equal atmospheric pressures. The air motion, or wind direction, is at right angles to the isobars and moves from a region of high pressure to a region of low pressure. From Spellman and Whiting (2006).

Localized air circulation gives rise to thermal circulation (a result of the relationship based on a law of physics whereby the pressure and volume of a gas are directly related to its temperature). A change in temperature causes a change in the pressure and volume of a gas. With a change in volume comes a change in density, since $P = m/V$, so regions of the atmosphere with different temperatures may have different air pressures and densities. As a result, localized heating sets up air motion and gives rise to thermal circulation. To gain understanding of this phenomenon, consider Figure 9.2.

Once the air has been set into motion, secondary forces (velocity-dependent forces) act. These secondary forces are (1) earth's rotation (Coriolis force) and (2) contact with the rotating earth (friction). The *Coriolis force*, named after its discoverer, French mathematician Gaspard Coriolis (1772–1843), is the effect of rotation on the atmosphere and on all objects on the earth's surface. In the northern hemisphere, it causes moving objects and currents to be deflected to the right; in the southern hemisphere, it causes deflection to the left, because of the earth's rotation. Air, in large-scale north or south movements, appears to be deflected from its expected path. That is, air moving pole-ward in the northern hemisphere appears to

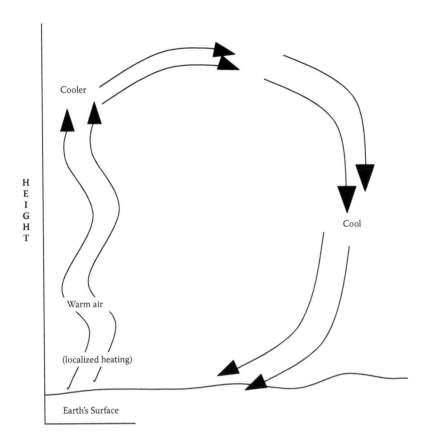

FIGURE 9.2 Thermal circulation of air. Localized heating, which causes air in the region to rise, initiates the circulation. As the warm air rises and cools, cool air near the surface moves horizontally into the region vacated by the rising air. The upper, still cooler, air then descends to occupy the region vacated by the cool air. From Spellman and Whiting (2006).

be deflected toward the east; air moving southward appears to be deflected toward the west.

Figure 9.3 illustrates the Coriolis effect on a propelled particle (analogous to the apparent effect of an air mass flowing from point A to point B). From Figure 9.3, the action of the earth's rotation on the air particle as it travels north over the earth's surface, as earth rotates beneath it from east to west, can be seen. Projected from point A to point B, the particle will actually reach point B because as it is moving in a straight line (deflected), the earth rotates east to west beneath it.

Friction (drag) can also cause the deflection of air movements. This friction (resistance) is both internal and external. The friction of its molecules generates internal friction. Friction is also generated when air molecules run into each other. External friction is caused by contact with terrestrial surfaces. The magnitude of the frictional force along a surface is dependent on the air's magnitude and speed, and the opposing frictional force is in the opposite direction of the air motion.

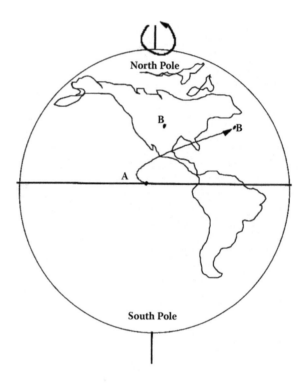

FIGURE 9.3 The effect of the earth's rotation on the trajectory of a propelled particle. From Spellman and Whiting (2006).

Did You Know?

Friction, one of the major forces affecting the wind, comes into play near the earth's surface and continues to be a factor up to altitudes of about 500 to 1,000 meters. This section of the atmosphere is referred to as the planetary or atmospheric boundary layer. Above this layer, friction no longer influences the wind.

LOCAL AND WORLD AIR CIRCULATION

Air moves in all directions, and these movements are essential for those of us on earth: Vertical air motion is essential in cloud formation and precipitation. Horizontal air movement near the earth's surface produces winds.

Wind is an important factor in human comfort, especially affecting how cold we feel. A brisk wind at moderately low temperatures can quickly make us uncomfortably cold. Wind promotes the loss of body heat, which aggravates the chilling effect, expressed through windchill factors in the winter (see Table 9.2) and the heat index in the summer (see Table 9.3). These two scales describe the cooling effects of wind on exposed flesh at various temperatures.

TABLE 9.2
Windchill Chart

	Temperature (°F)											
Wind mph	30	25	20	15	10	5	0	−5	−10	−15	−20	−25
5	25	19	13	7	1	−5	−11	−16	−22	−28	−34	−40
10	21	15	9	3	−4	−10	−16	−22	−28	−35	−41	−47
15	19	13	3	0	−7	−13	−19	−26	−32	−39	−45	−51
20	17	11	4	−2	−9	−15	−22	−29	−35	−42	−48	−55
25	16	9	3	−4	−11	−17	−24	−31	−37	−44	−51	−58
30	15	8	1	−5	−12	−19	−26	−33	−39	−46	−53	−60
35	14	7	0	−7	−14	−21	−27	−34	−41	−48	−55	−62
40	13	6	−1	−8	−15	−22	−29	−36	−43	−50	−57	−64
45	12	5	−2	−9	−16	−23	−30	−37	−44	−51	−58	−65
50	12	4	−3	−19	−17	−24	−31	−38	−45	−52	−60	−67
55	11	4	−3	−11	−18	−25	−32	−39	−46	−54	−61	−68
60	10	3	−4	−11	−19	−26	−33	−40	−48	−55	−62	−69

Source: USA Today, http://www.usatoday.com/weather/resources/basics/windchill/wind-chill-chart.htm.

TABLE 9.3
Heat Index Chart (Temperature and Relative Humidity)

	Temperature (°F)															
RH (%)	90	91	92	93	94	95	96	97	98	99	100	101	102	103	104	105
90	119	123	128	132	137	141	146	152	157	163	168	174	180	186	193	199
85	115	119	123	127	132	136	141	145	150	155	161	166	172	178	184	190
80	112	115	119	123	127	131	135	140	144	149	154	159	164	169	175	180
75	109	112	115	119	122	126	130	134	138	143	147	152	156	161	166	171
70	106	109	112	115	118	122	125	129	133	137	141	145	149	154	158	163
65	103	106	108	111	114	117	121	124	127	131	135	139	143	147	151	155
60	100	103	105	108	111	114	116	120	123	126	129	133	136	140	144	148
55	98	100	103	105	107	110	113	115	118	121	124	127	131	134	137	141
50	96	98	100	102	104	107	109	112	114	117	119	122	125	128	131	135
45	94	96	98	100	102	104	106	108	110	113	115	118	120	123	126	129
40	92	94	96	97	99	101	103	105	107	109	111	113	116	118	121	123
35	91	92	94	95	97	98	100	102	104	106	107	109	112	114	116	118
30	89	90	92	93	95	96	98	99	101	102	104	106	108	110	112	114

Source: Weather images: http://www.weatherimages.org/data/heatindex.html.

Note: Exposure to full sunshine can increase HI values by up to 15°F.

Local winds are the result of atmospheric pressure differences involved with thermal circulations because of geographic features. Land areas heat up more quickly than do water areas, giving rise to a convection cycle. As a result, during the day, when land is warmer than the water, we experience a lake or sea breeze.

At night, the cycle reverses. Land loses its heat more quickly than water, so the air over the water is warmer. The convection cycle sets to work in the opposite direction and a land breeze blows.

In the upper troposphere (11 to 14 kilometers, west to east flows) are very narrow fast-moving bands of air called jet streams. Jet streams have significant effects on surface airflows. When jet streams accelerate, divergence of air occurs at that altitude. This promotes convergence near the surface and the formation of cyclonic motion. Deceleration causes convergency aloft and subsidence near the surface, causing an intensification of high-pressure systems.

Jet streams are thought to result from the general circulation structure in the regions where great high- and low-pressure areas meet.

Blowing in the Wind

Wind energy conversion—the leading mechanically based renewable energy for much of human history—has been around for thousands of years. It is a technology that has been reinvented numerous times, a proven technology with many advantages for the consumer. Modern wind farms demonstrate that wind turbines are a viable alternative to fossil fuel energy production.

Since cost and capacity factors are so different between wind-generated energy and fossil-fuel-generated energy, low installed-cost-per-kilowatt figures for wind turbines are somewhat misleading because of the low capacity factor of wind turbines relative to coal and other fossil-fueled power plants. (Note: Capacity factor is the ratio of actual energy produced by a power plant to the potential energy produced if the plant operated at rated capacity for a full year.) Capacity factors of successful wind farm operations range from 0.20 to 0.35. Fossil fuel power plants have factors of more than 0.50, and some of the new gas turbines reach over 0.60.

Capacity factor and the difference between low- and high-capacity production are also misleading. Wind conversion capacity is flexible; production levels vary with the density of the wind resource. More importantly, the wind resource is constant for the life of the machine—not subject to cost manipulation or cost increases. Fossil fuels as energy sources are popular with investors because many of the risks are passed on to consumers. When fossil fuel shortages occur, the investors can raise their prices—causing an *increase* in revenues for investors. In a nasty twist for the consumer, investors in fossil fuel energy production are *rewarded* for (1) speeding the depletion of a nonrenewable resource or (2) not investing enough of their profits to support infrastructure, which drives up prices (think California in 2000–2001). This seeming advantage for wind conversion technology thus becomes a barrier to investment: if big oil, coal, or gas companies could charge consumers for the wind, wind power development would have been a done deal long ago.

The cost of energy from larger electrical output wind turbines used in utility-interconnected or wind farm applications dropped from more than $1.00 per kilowatt-hour (kWh) in 1978 to under $0.05 per kWh in 1998, and dropped to $0.025 per kWh when new large wind plants came on line in 2001 and 2002. Hardware costs have dropped below $800 per installed kilowatt, lower than the capital costs of almost every other type of power plant.

Wind energy soon will be the most cost effective source of electrical power, and perhaps has already achieved this status. The actual life cycle cost of fossil fuels (starting with coal mining and fuel extraction, including transport and use technology, and factoring in environmental impact and political costs) is not really known, but is certainly far higher than the current wholesale rates—and has been loaded squarely on the shoulders of consumers by the energy industry. From strictly a fuel-cost perspective, since fossil fuel resources are nonrenewable, the eventual depletion of these energy sources will entail rapid escalations in price. Add to this the environmental and political costs of fossil fuel use, and the increased awareness of the public to these issues, and fossil fuel become even more expensive.

Wind energy experts are hopeful for the future of their industry. While infinite refinements and improvements are possible, the major technology developments that allow commercialization are complete. "At some point, a 'weather change' in the marketplace, or a 'killer application' somewhere will put several key companies or financial organizations in a position to profit, and wind energy conversion investors will take advantage of public interest, the political and economic climate, and emotional or marketing factors to position wind energy technology (developed in a long lineage from the Chinese and the Persians to the present wind energy researchers and developers) for its next round of development" (Spellman and Whiting, 2006).

Though wind energy production is generally considered unusually environmentally clean, serious environmental issues do exist. For species protection, wind farm placement should be carefully studied. Wind farms put stresses on already fragmented and reduced wildlife habitats. Another serious factor is the avian mortality rate. Just as high-rise buildings, power lines, towers, antennas, and other man-made structures are passive killers of many birds, badly positioned wind farms put a heavy toll on bird populations, especially upon migratory birds. The Altamont Pass wind farms (near San Francisco) are badly placed, and since their construction in the 1980s, have killed many golden eagles and other species as well.

Golden eagles lock on to a prey animal and dive for it, totally blocking out the threat of the wind turbine. They can see the propellers under normal circumstances, but their instinctive prey focus is so strong that when they stoop over a kill, they see only their prey.

Six to ten different companies, including U.S. Wind Power, Kenetech Wind Power, and Green Mountain Energy, own the turbines at the Altamont Pass wind farms—over 7,000 of them. Another wind facility in Tehachapi Pass near Los Angeles poses little threat to bird populations.

In an interview with a reporter from the *San Francisco Chronicle*, conservationist Stan Moore states: "It is estimated that 40 to 60 golden eagles are killed annually, plus 200 red-tailed hawks and smaller numbers of American kestrels, crows, burrowing owls and other birds. Those numbers are conservative. . . .

"I'm in favor of renewable energy when it is sited appropriately, but Altamont Pass is one of the worst places to put a wind farm on planet Earth, because it is adjacent to one of the densest breeding populations of golden eagles in the world. It's a unique place for raptors because of the abundant food source in ground squirrels. . . .

"Altamont Pass is not an appropriate place for wind turbines. What we have there is world-class golden-eagle habitat."

The California Energy Commission financed a 5-year study in 1994, conducted by Dr. Grainger Hunt, a world authority on birds of prey who works with the Santa Cruz Predatory Bird Research Group. The study detected no population-level impacts for golden eagles in Altamont Pass; however, the local eagles could provide source population for all of California, if the wind farms deaths were halted. Instead, the local eagles are an at-risk population: if other pressures disturbed the Altamont Pass golden eagle population—an outbreak of West Nile virus, for example—catastrophic population losses would occur, because the wind turbines have removed much of the buffer population.

Because control guidelines are voluntary, not mandatory, the energy industry essentially polices itself on this issue. When the U.S. Fish and Wildlife Service (practicing what services officials themselves call discretionary law enforcement of service laws) chooses not to enforce the Migratory Bird Treaty Act and the Bald and Golden Eagle Protection Act, and when California officials fail to enforce their own decrees (a state designation of the golden eagle as a "fully protected species" and a "species of special concern"), the protections supposedly provided by federal and state laws become a farce. (Hank Pellissier, "Golden Eagle Eco-Atrocity at Altamont Pass," special to SF Gate: *San Francisco Chronicle*, 2003; Xcel Energy Ponnequin wind farm in northeastern Colorado; http://telosnet.com/wind/ (accessed January 7, 2008)).

REFERENCES AND RECOMMENDED READING

Anthes, R. A. 1996. *Meteorology.* 7th ed. Upper Saddle River, NJ: Prentice Hall.

Anthes, R. A., Cahir, J. J., Fraizer, A. B., and Panofsky, H. A. 1984. *The Atmosphere.* 3rd ed. Columbus, OH: Charles E. Merrill Publishing Company.

Ingersoll, A. P. 1983. The atmosphere. *Scientific American* 249:162–74.

Lutgens, F. K., and Tarbuck, E. J. 1982. *The atmosphere, an introduction to meteoro-logy.* Englewood Cliffs, NJ: Prentice-Hall.

Miller, G. R., Jr. 2004. *Environmental science.* 10th ed. Sydney, Australia: Thompson-Brooks/Cole.

Moron, J. M., Morgan, M. D., and Wiersma, J. H. 1986. *Introduction to environmental science.* 2nd ed. New York: W.H. Freeman & Company.

Shipman, J. T., Adams, J. L., and Wilson, J. D. 1987. *An introduction to physical science.* 5th ed. Lexington, MA: D.C. Heath & Company.

Spellman, F. R., and Whiting, N. E. 2006. *Environmental science and technology: Concepts and applications.* 2nd ed. Rockville, MD: CRC Press.

10 Weather and Climate

The Pharisees also with the Sadducees came, and tempting desired him that he would show them a sign from heaven. He answered and said unto them, When it is evening ye say, It will be fair weather today for the sky is red and lowering. Oh ye hypocrites, ye can discern the face of the sky, but can ye not discern the signs of the times. (Matthew 16:1–4)

Mean Weather

Intermittent rain, I've learned,
Which forecasts tell about,
Is rain that stops when I go in
And starts when I come out.

Elizabeth Dolan, in *The Breeze*, vol. 2, no. 8, September 10, 1945, p. 6

INTRODUCTION

An eminent meteorologist once said, "A butterfly flapping its wings in Brazil can cause a tornado in Texas." What the meteorologist was implying is true to a point (and in line with what some critics might say): because of tiny nuances in earth's weather patterns, making accurate, long-range weather predictions is extremely difficult.

What is the difference between weather and climate? Some people get these two confused, believing they mean the same thing, but they do not. In this chapter you will gain a clear understanding of the meaning of and difference between the two, as well as an understanding of the role weather plays in air pollution.

METEOROLOGY: THE SCIENCE OF WEATHER

Meteorology is the science concerned with the atmosphere and its phenomena. The atmosphere is the media into which all air pollution is emitted. The meteorologist observes atmospheric processes such as temperature, density, (air) winds, clouds, precipitation, and other characteristics and endeavors to account for their observed structure and evaluation (weather, in part) in terms of external influence and the basic laws of physics. *Air pollution meteorology* is the study of how these atmospheric processes affect the fate of air pollutants.

Since the atmosphere serves as the medium into which air pollutants are released, the transport and dispersion of these releases are influenced significantly by meteorological parameters. Understanding air pollution meteorology and its influence in pollutant dispersion is essential in air quality planning activities.

Planners use this knowledge to help locate air pollution monitoring stations and develop implementation plans to bring ambient air quality into compliance with standards. Meteorology is used in predicting the ambient impact of a new source of air pollution and to determine the effect on air quality from modifications to existing sources (EPA, 2005).

Weather is the state of the atmosphere, mainly with respect to its effect upon life and human activities; as distinguished from *climate* (the long-term manifestations of weather), weather consists of the short-term (minutes or months) variations of the atmosphere. Weather is defined primarily in terms of heat, pressure, wind, and moisture.

At high levels above the earth, where the atmosphere thins to near vacuum, there is no weather; instead, weather is a near-surface phenomenon. This is evidenced clearly on a day-by-day basis where you see the ever-changing, sometimes dramatic, and often violent weather display.

In the study of air science, and in particular of air quality, the following determining factors are directly related to the dynamics of the atmosphere, resulting in local weather. These factors include strength of winds, the direction they are blowing, temperature, available sunlight (needed to trigger photochemical reactions, which produce smog), and the length of time since the last weather event (strong winds and heavy precipitation) cleared the air.

Weather events (such as strong winds and heavy precipitation) that work to clean the air we breathe are beneficial, obviously. However, few people would categorize the weather events such as tornadoes, hurricanes, and typhoons as beneficial. Other weather events have both a positive and a negative effect. One such event is El Nino–Southern Oscillation, discussed below.

EL NINO–SOUTHERN OSCILLATION

El Nino–Southern Oscillation (ENSO) is a natural phenomenon that occurs every 2 to 9 years on an irregular and unpredictable basis. El Nino is a warming of the surface waters in the tropical eastern Pacific, which causes fish to disperse to cooler waters and, in turn, causes the adult birds to fly off in search of new food sources elsewhere.

Through a complex web of events, El Nino (which means "the child" in Spanish because it usually occurs during the Christmas season off the coasts of Peru and Ecuador) can have a devastating impact on all forms of marine life.

During a normal year, equatorial trade winds pile up warm surface waters in the western Pacific. Thunderheads unleash heat and torrents of rain. This heightens the east–west temperature difference, sustaining the cycle. The jet stream blows from North Asia to California. During an El Nino–Southern Oscillation year, trade winds weaken, allowing warm waters to move east. This decreases the east–west temperature difference. The jet stream is pulled farther south than normal, picks up storms it would usually miss, and carries them to Canada or California. Warm waters eventually reach South America.

One of the first signs of its appearance is a shifting of winds along the equator in the Pacific Ocean. The normal easterly winds reverse direction and drag a large mass of warm water eastward toward the South American coastline. The large mass of warm water basically forms a barrier that prevents the upwelling of nutrient-rich cold water from the ocean bottom to the surface. As a result, the growth of microscopic algae that normally flourish in the nutrient-rich upwelling areas diminishes sharply, and that decrease has further repercussions. For example, El Nino–Southern Oscillation has been linked to patterns of subsequent droughts, floods, typhoons, and other costly weather extremes around the globe. Take a look at El Nino–Southern Oscillation's effect on the West Coast of the United States, where ENSO has been blamed for West Coast hurricanes, floods, and early snowstorms. On the positive side, ENSO typically brings good news to those who live on the East Coast of the United States: a reduction in the number and severity of hurricanes.

Note that in addition to reducing the number and severity of hurricanes, in October 1997 the Associated Press reported that a new study has shown that ENSO also deserves credit for invigorating plants and helping to control the pollutant linked to global warming. Researchers have found that El Nino causes a burst of plant growth throughout the world, and this removes carbon dioxide from the atmosphere.

Atmospheric carbon dioxide (CO_2) has been increasing steadily for decades. The culprits are increased use of fossil fuels and the clearing of tropical rainforests. However, during an ENSO phenomenon, global weather is warmer, there is an increase in new plant growth, and CO_2 levels decrease.

Not only does ENSO have a major regional impact in the Pacific, but its influence extends to other parts of the world through the interaction of pressure, airflow, and temperature effects.

El Nino–Southern Oscillation is a phenomenon that, although not quite yet completely understood by scientists, causes both positive and negative results, depending upon where you live.

THE SUN: THE WEATHER GENERATOR

The sun is the driving force behind weather. Without the distribution and reradiation to space of solar energy, we would experience no weather (as we know it) on earth. The sun is the source of most of the earth's heat. Of the gigantic amount of solar energy generated by the sun, only a small portion bombards earth. Most of the sun's solar energy is lost in space. A little over 40% of the sun's radiation reaching earth hits the surface and is changed to heat. The rest stays in the atmosphere or is reflected back into space.

Like a greenhouse, the earth's atmosphere admits most of the solar radiation. When solar radiation is absorbed by the earth's surface, it is reradiated as heat waves, most of which are trapped by carbon dioxide and water vapor in the atmosphere, which work to keep the earth warm in the same way a greenhouse traps heat.

By now you are aware of the many functions performed by the earth's atmosphere. You should also know that the atmosphere plays an important role in regulating the earth's heating supply. The atmosphere protects the earth from too much solar radiation during the day and prevents most of the heat from escaping at night. Without the filtering and insulating properties of the atmosphere, the earth would experience severe temperatures, similar to other planets.

On bright clear nights the earth cools more rapidly than on cloudy nights because cloud cover reflects a large amount of heat back to earth, where it is reabsorbed.

The earth's air is heated primarily by contact with the warm earth. When air is warmed, it expands and becomes lighter. Air warmed by contact with earth rises and is replaced by cold air, which flows in and under it. When this cold air is warmed, it too rises and is replaced by cold air. This cycle continues and generates a circulation of warm and cold air, which is called *convection*.

At the earth's equator, the air receives much more heat than the air at the poles. This warm air at the equator is replaced by colder air flowing in from north and south. The warm, light air rises and moves poleward high above the earth. As it cools, it sinks, replacing the cool surface air, which has moved toward the equator.

The circulating movement of warm and cold air (convection) and the differences in heating cause local winds and breezes. Different amounts of heat are absorbed by different land and water surfaces. Soil that is dark and freshly plowed absorbs much more than grassy fields. Land warms faster than does water during the day and cools faster at night. Consequently, the air above such surfaces is warmed and cooled, resulting in production of local winds.

Winds should not be confused with air currents. Wind is primarily oriented toward horizontal flow. Air currents, on the other hand, are created by air moving upward and downward. Wind and air currents have direct impact on air pollution. Air pollutants are carried and dispersed by wind. An important factor in determining the areas most affected by an air pollution source is wind direction. Since air pollution is a global problem, wind direction on a global scale is important.

Along with wind, another constituent associated with the earth's atmosphere is water. Water is always present in the air. It evaporates from the earth, two thirds of which is covered by water. In the air, water exists in three states: solid, liquid, and invisible vapor.

The amount of water in the air is called humidity. The *relative humidity* is the ratio of the actual amount of moisture in the air to the amount needed for saturation at the same temperature. Warm air can hold more water than cold. When air with a given amount of water vapor cools, its relative humidity increases; when the air is warmed, its relative humidity decreases.

AIR MASSES

An air mass is a vast body of air (a macroscale phenomenon that can have global implications) in which the conditions of temperature and moisture are much the same at all points in a horizontal direction. An air mass takes on the temperature and moisture characteristics of the surface over which it forms and travels, though its original characteristics tend to persist. The processes of radiation, convection, condensation, and evaporation condition the air in an air mass as it travels. Also, pollutants released into an air mass travel and disperse within the air mass. Air masses develop more commonly in some regions than in others. Table 10.1 summarizes air masses and their properties.

When two different air masses collide, a *front* is formed. A front is not a sharp wall but a zone of transition that is often several miles wide. Four frontal patterns—warm,

TABLE 10.1
Classification of Air Masses

Name	Origin	Properties	Symbol
Artic	Polar regions	Low temperatures; low specific but high summer relative humidity; the coldest of the winter air masses	A
Polar continental[a]	Subpolar continental areas	Low temperatures (increasing with southward movement); low humidity, remaining constant	cP
Polar maritime	Subpolar area and arctic region	Low temperatures increasing with movement, higher humidity	mP
Tropical continental	Subtropical high-pressure land areas	High temperatures, low moisture content	cT
Tropical maritime	Southern borders of oceanic subtropical, high-pressure areas	Moderate high temperatures, high relative and specific humidity	mT

Source: EPA (2005).

[a] The name of an air mass, such as polar continental, can be reversed to continental polar, but the symbol, cP, is the same for either name.

cold, occluded, and stationary—can be formed by air of different temperatures. A *cold front* marks the line of advance of a cold air mass from below, as it displaces a warm air mass. A *warm front* marks the advance of a warm air mass as it rises up over a cold one.

When cold and warm fronts merge (the cold front overtaking the warm front), *occluded fronts* form. Occluded fronts can be called cold front or warm front occlusions. But in either case, a colder air mass takes over an air mass that is not as cold.

The last type of front is the *stationary front*. As the name implies, the air masses around this front are not in motion. A stationary front can cause bad weather conditions that persist for several days.

THERMAL INVERSIONS AND AIR POLLUTION

Earlier, it was pointed out that during the day the sun warms the air near the earth's surface. Normally, this heated air expands and rises during the day, diluting low-lying pollutants and carrying them higher into the atmosphere. Air from surrounding high-pressure areas then moves down into the low-pressure area created when the hot air rises (see Figure 10.1). This continual mixing of the air helps keep pollutants from reaching dangerous levels in the air near the ground.

Sometimes, however, a layer of dense, cool air is trapped beneath a layer of less dense, warm air in a valley or urban basin. This is called a *thermal inversion*. In effect, a warm-air lid covers the region and prevents pollutants from escaping in upward-flowing air currents. Usually these inversions trap air pollutants (i.e., plume dispersion is inhibited) at ground level for a short period of time. However, sometimes

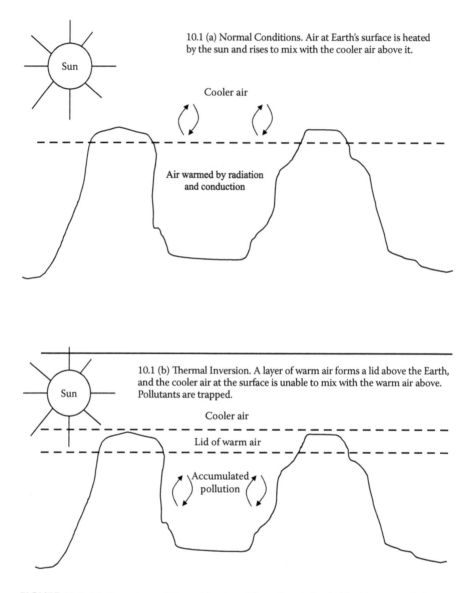

10.1 (a) Normal Conditions. Air at Earth's surface is heated by the sun and rises to mix with the cooler air above it.

Cooler air

Air warmed by radiation and conduction

10.1 (b) Thermal Inversion. A layer of warm air forms a lid above the Earth, and the cooler air at the surface is unable to mix with the warm air above. Pollutants are trapped.

Cooler air

Lid of warm air

Accumulated pollution

FIGURE 10.1 (a) Normal conditions. Air at earth's surface is heated by the sun and rises to mix with the cooler air above it. (b) Thermal inversion. A layer of warm air forms a lid above the earth, and the cooler air at the surface is unable to mix with the warm air above. Pollutants are trapped.

they last for several days when a high-pressure air mass stalls over an area, trapping air pollutants at ground level, where they accumulate to dangerous levels.

The best known location in the United States where thermal inversions occur on an almost daily basis is in the Los Angeles Basin. The Los Angeles Basin is a valley with a warm climate and light winds, surrounded by mountains, located near the

Pacific Coast. Los Angeles is a large city with a large population of people and automobiles and possesses the ideal conditions for smog, which is worsened by frequent thermal inversions.

REFERENCES AND RECOMMENDED READING

EPA. 2005. Basic air pollution meteorology. APTI Course SI 409. www.epa.gov/apt (accessed January 8, 2008).

Spellman, F. R., and Whiting, N. 2006. *Environmental science and technology: Concepts and applications.* 2nd ed. Rockville, MD: Government Institutes.

11 Microclimates

Nothing that is can pause or stay;
The moon will wax, the moon will wane,
The mist and cloud will turn to rain,
The rain to mist and cloud again,
Tomorrow be today.

Henry Wadsworth Longfellow

INTRODUCTION

When we think about climate, we are generally referring to overall or generalized weather conditions at a particular place or region over a period of time. In addition to precipitation and temperature, climates have been classified into zones by vegetation, moisture index, and even measures of human discomfort. Using the general climate zone names allows us to differentiate between a particular climate (with its specific climatic conditions) and another one with differing conditions. When geographical patterns in the weather occur again and again over a long period, they can all be used to define the climate of a region. Some climate zones are known as the hot climates, which include desert, tropical continental, tropical monsoon, tropical marine, or equatorial types. Warm climates include west coast (Mediterranean) and warm east coast. Another category includes the cool climates, such as cold desert, west coast (cool), cool temperature interior, and cool temperate east coast types. Finally, there are the mountain and the cold climate categories of cold continental and polar or tundra. Each of these different climate types is differentiated from the others. However, they all have one major feature in common: they are large-scale regional climates (with variations), occurring at various places throughout the world. They only consider the broad similarities between a particular climate at various locations worldwide; local differences are ignored and boundaries are approximate.

What factors determine the variations of climate over the surface of the earth? The primary factors are:

1. The effect of latitude and the tilt of the earth's axis to the plane of the orbit about the sun
2. The large-scale movements of different wind belts over the earth's surface
3. The temperature difference between land and sea
4. The contours of the earth's surface
5. The location of the area in relation to ocean currents

What factors determine how climates are distributed? When considering climate distribution, remember that the world does not fall into compartments. The globe is a mosaic of many different types of climate, just as it is a mosaic of numerous types of ecosystems. The complexity of the distribution of land and sea and the consequent complexity of the general circulation of the atmosphere have a direct effect on the distribution of the climate.

MICROCLIMATES

What is a microclimate? In answering this question, we must first discuss scale. For example, let us take a look at flow of air within a very small environment: the emission of smoke from a chimney. This flow represents one of the smallest spatial subdivisions of atmospheric motion, or microscale weather. On a more realistic but still relatively small scale, we must consider the geographical, biological, and man-made features that make local climate different from the general climate. This local climatic pattern is called a *microclimate*.

What are the elements or conditions that cause local or microclimates? Location and local conditions are the main ingredients making up a microclimate. Let us look at one example.

Large inland lakes moderate temperature extremes and climatic differences between the windward and lee sides. For example, Seattle, on the windward side of Lake Washington, and Bellevue, on the lee side, only about nine miles east, have microclimatic differences (although modest) between the two cities. These microclimatic differences include temperature fluctuations, precipitation levels, wind speed, and relative humidity.

Even more dramatic differences can be seen in such parameters when a comparison is made between a city such as Milwaukee, on the windward side of Lake Michigan, and Grand Haven, on the lee side, only 85 miles east.

Other examples of microclimates are listed below:

- Near the ground
- Over open land areas
- In woodlands or forested areas
- In valley regions
- In hillside regions
- In urban areas
- In seaside locations

In the following sections we take a closer look at these microclimates: at their nature, causative factors, and geographical/topographical locations.

MICROCLIMATES NEAR THE GROUND

Nowhere in the atmosphere are climatic differences as distinct as they are near the ground. For instance, when you go to the beach on a warm summer day, you no doubt notice that the grass and water are much cooler to your feet than the sand. So, you may ask, what is it about this area near the ground that produces a microclimate with such major differences?

It is the interface (or activity zone) between the atmosphere and the ground surface (sandy shore) that causes the stark difference in temperature variability. Energy is reaching the sandy beach from the sun and from the atmosphere (though to a much lesser extent). The energy is either reflected and then returned to the atmosphere in a different form, or is absorbed and stored in the sandy surface as heat.

Ground-level energy absorption is very sensitive to the nature of the ground surface. Ground surface color, wetness, cover (vegetation), and topography are conditions that all affect the interaction between the ground and the atmosphere. Consider a snow-covered ground, for example. Clean snow reflects solar radiation, so the surface remains cool and the snow fails to melt. However, dirty snow absorbs more radiation, heats up, and is likely to melt. If the snowy area is shielded by vegetation, the vegetation, too, may protect the snow from the heat of the sun.

A surface cover such as clean snow has the ability to reflect solar radiation because of its high albedo. *Albedo* (the ratio between the light reflected from a surface and the total light falling on it) always has a value less than or equal to 1. An object with a high albedo, near 1, is very bright, while a body with a low albedo, near 0, is dark. For example, freshly fallen snow typically has an albedo that is between 75 and 90%; that is, 75 to 95% of the solar radiation that is incident on snow is reflected. At the other extreme, the albedo of a rough, dark surface, such as a green forest, may be as low as 5%. The albedos of some common surfaces are listed in Table 11.1. The portion of insolation not reflected is absorbed by the earth's surface, warming it. This means earth's albedo plays an important part in the earth's radiation balance and influences the mean annual temperature and the climate on both local and global scales.

TABLE 11.1
The Albedo of Some Surface Types in Percent Reflected

Surface	Albedo
Water (low sun)	10–100
Water (high sun)	3–10
Grass	16–26
Glacier ice	20–40
Deciduous forest	15–20
Coniferous forest	5–15
Old snow	40–70
Fresh snow	75–95
Sea ice	30–40
Blacktopped tarmac	5–10
Desert	25–30
Crops	15–25

Microclimate over Open Land Areas

Many different properties of ground layer or soil type influence conditions in the thin layer of atmosphere just above it. Light-colored soils do not absorb energy as efficiently as do organically rich darker soils. Another important factor is soil moisture. Wet soils are normally dark, but moist soil (because water has a large heat capacity) requires a great deal of energy to raise its temperature. A moist soil warms up more slowly than a dry one.

Soil is a heterogeneous mixture of various particles. In between the soil particles is a large amount of air—air that is a poor conductor of heat. The larger the amount of air between the soil particles, the slower the heat transfers through the soil. As demonstrated in our example of the sandy beach, on a hot sunny day the heat is trapped in the upper layers, so the surface layers warm up more rapidly and become extremely hot. Water conducts heat more readily than air, so soils that contain some moisture are able to transmit warmth away from the surface more easily than dry soils. This is not always the case, however. If the soil contains too much water, the large heat capacity of the water will prevent the soil from warming despite heat being conducted from the surface.

Microclimates in Woodlands or Forested Areas

When making microclimate comparisons between open land areas and forested areas (commonly referred to as a forest climate), the differences are quite apparent. Forested areas, for example, are generally warmer in winter than the open areas, while open land is warmer in summer than forested areas. The forest climate has reduced wind speeds, while the open land area has higher wind speeds. The forest climate has higher relative humidity, while the open area has lower relative humidity. In the forest climate, water storage capacity is higher and evaporation rates are lower, while in the open land area water storage capacity is lower and evaporation is higher.

Microclimates in Valley and Hillside Regions

Heavy, cold air flows downhill, forming cold pockets in valleys. Frost is much more common there, so orchards of apples and oranges and vines of grapes are planted on hillsides to ensure frost drainage when cold spells come.

Probably the best way in which to describe the microclimate in a typical valley region is to compare and contrast it with a hillside environment.

In a typical valley region, the daily minimum temperature is much lower than that in a hillside area. The daily and annual temperature range for a valley is much larger than that of a hillside area. In a valley region, more frost occurs than in a hillside region. Wind speed at night is lower in a valley than on a hillside, and morning fog is more prevalent and lasts longer in a valley region.

Microclimates in Urban Areas

The microclimate in an urban area, compared to that of the countryside, is usually quite obvious. A city, for example, is usually characterized by having haze and smog, higher temperatures, lower wind speed, and reduced radiation. The countryside, on the other hand, is characterized by clear air, lower temperatures, and high wind speeds and radiation.

These different microclimatic conditions should come as no surprise to anyone, especially when you consider what happens when a city is built. Instead of a mixture of soil or vegetation, the surface layer is covered with concrete, brick, glass, and stone surfaces ranging to heights of several hundred feet. These materials have vastly different physical properties from soil and trees. They shed and carry away water, absorb heat, block and channel the passage of winds, and present albedo levels significantly different from those of the natural world. All of these factors (and more) work to alter the climate conditions in the area.

Did You Know?

Urban areas have added roughness features and different thermal characteristics due to the presence of man-made elements. The thermal influence dominates the influence of the frictional components. Building materials such as brick and concrete absorb and hold heat more efficiently than soil and vegetation found in rural areas. After the sun sets, the urban area continues to radiate heat from buildings, paved surfaces, etc. Air warmed by the urban complex rises to create a dome over the city. It is called the *heat island effect*. The city emits heat all night. Just when the urban area begins to cool, the sun rises and heats the urban complex again. Generally, city areas never revert to stable conditions because of the continual heating that occurs (EPA, 2005).

MICROCLIMATES IN SEASIDE LOCATIONS

The major climatic feature associated with seaside locations is the sea breeze. Sea breezes are formed by the different responses to heating of water and land. For example, if we have a bright, sunny morning with little wind, the ground surface warms rapidly as it absorbs short-wave radiation. Most of this heat is retained at the surface, although some will be transferred through the soil. As a result, the temperature of the ground surface increases and some of the heat warms the air above. When the sun sets, the surface starts to cool rapidly, because there is little store of heat in the soil. Thus, we find that land surfaces are characterized by high day (and summer) temperatures and low night (and winter) temperatures.

Now let us take a look at the response of the sea, which is very different. Solar energy (sunshine) is able to penetrate through the water to a certain level. Much solar energy has to be absorbed to raise its temperature. Through wave action and convection, the warm surface water is mixed with cooler deeper water. With enough solar energy and time, the top several feet of water forms an active layer where temperature change is slow. Slight warming occurs during the day and slight cooling at night. This means that the sea is normally cooler than the land by day and warmer by night.

The higher temperature over the land by day generates a weak low-pressure area. As this intensifies during daytime heating, a flow of cool, more humid air spreads inland from the sea, gradually changing in strength and direction during the day. At night the reverse occurs, with circulation of air from the cooler land to the warmer sea, though as the temperature difference is usually less, the land breeze is weak. Even large lakes can show a breeze system of this nature.

REFERENCES AND RECOMMENDED READING

EPA. 2005. Basic air pollution meteorology. APTI Course SI:409. www.epa.gov/apti (accessed January 9, 2008).

Spellman, F. R., and Whiting, N. 2006. *Environmental science and technology: Concepts and applications.* 2nd ed. Rockville, MD: Government Institutes.

12 The Endangered Atmosphere
Climate Change

Humanity is conducting an unintended, uncontrolled, globally pervasive experiment whose ultimate consequences could be second only to nuclear war. The earth's atmosphere is being changed at an unprecedented rate by pollutants resulting from human activities, inefficient and wasteful fossil fuel use, and the effects of rapid population growth in many regions. These changes are already having harmful consequences over many parts of the globe.

Toronto Conference statement, June 1988

SETTING THE STAGE

Time: 10,312 BCE

He sat on the ground leaning against a deadfall, his leather-wrapped legs pulled tight under him, and watched the swamp. He felt disoriented, detached from the world around him. Even the air around him felt strange—it was different; this place was unusually warm. A possum waddled from a copse of vine maple below him, and Yurk watched the possum move off to the left—the possum, in a hurry, constantly jerking his head to the right, over his shoulder. The possum darted toward the marshy bank, and stopped to sniff the ground. Some noise, or an odor carried on the wind, seemed suddenly to startle the possum into attention, and he looked back toward Yurk, then moved off into the tall marsh grass, where he disappeared from view.

The rain was coming down in a fine drizzle. The wind sighed through the fir boughs and the afternoon was redolent with the smell of tree-perfumed air. Even with the light rain and wind, though, Yurk was warm—warmer than he ever remembered being before. He had never been so warm, his whole body at the same time. By a fire, only what faces the fire is warm.

Yurk's weathered face wore a mesmerized look as he chewed on a piece of bark, resting against the decaying trunk of the fallen tree, almost as if he was unaware of his surroundings. His eyes glazed over, as if he was there in shell form only—an empty one at that. Maybe his blank state and hypnotized appearance were the result of the view in front of those blank eyes: great truncated tree trunks blackened by fire stood above the surface of the swamp water—stark remnants of a very ancient past. A misty pall hung over the swamp as the blackness lowered over the forest and the swamp took on an eerie, forbidding, spectral quality with the coming of night.

The cry of an owl drifted through the dark forest as Yurk stood (an effort that required much exertion from his tired, ancient body). Carefully he stretched and yawned—careful not because of his frailty, or from a sense of impending danger, but because of instinct—not fear exactly, just instinct. A lifetime, generations of lifetimes of vigilance for survival (both conscious and unconscious) had taught Yurk to be vigilant at all times in this place and time. He was leg-weary and footsore, but that really did not concern him. He knew his ending time was near—that was why he had traveled more than 200 miles to this place. This place he had come to had been familiar to him years before, but in a very different form. He wanted to see the wonderment of the swampy terrain that lay before him now.

Yurk was viewing something he had heard about from other clan members but something he had never witnessed before: a swamp.

Yes, a swamp, with blackened, truncated tree remnants. In all his years (unusually old for his time and circumstances, well more than sixty) Yurk had never seen such a sight. Before—up until now—the landscape he had been familiar with had been covered in snow and ice. He had visited this place many times in the past—what seemed a bare plain of ice and snow. He had not been on this journey in many years, but the last time he had come, he had simply trudged through the open area (the swamp) over a bridge of thick ice and snow. He (and no one else) had any idea that the swamp lay below the thick layers of ice and snow. In his absence from this place, he had heard the tales from the younger clan hunters, and had decided to take his last journey—to see such a place, such a site—before he died.

It was so warm.

As he stood, wiping his wet brow and looking out upon the swamp, 30 feet to Yurk's left, working toward the top of the steep, craggy ledge on the shear cliff edge, climbed the cat.

Like Yurk, the cat had come to this place many times in the past; although she could not cognitively determine the exact difference between the past and the present, she, too, knew this place had changed.

It was so warm.

In the past 15 or so years, the cat, along with her running mates (these cats almost always ventured into the wilderness accompanied—to hunt and to kill required help— sometimes lots of help), had, like Yurk and his clan members, crossed the swamp using the ice bridge. But now things were different; the cat knew this. She also knew that something else was different; it was so warm.

The cat (known today as Smilodon or saber tooth tiger) continued slowly, inexorably up the steep slope of the stony ridge. Unlike climbing to this ledge in the past, when she had allowed for the slipperiness of the ice sheet that covered the ledge, she should have had very little difficulty climbing the high terminal edge, overlooking the swamp. But now things were different—much different. She was on her last legs, in all ways. But her difficulty was even more than that; even though the going was easier now without the ice and snow, she still struggled her way up to the terminal point—it was so hot. She labored even to breathe.

Yurk and the cat were aware of each other. Each knew the other was there—have no doubt about that. Yurk probably more fearful of the cat than she was of him, but how could anyone tell? They had been bitter enemies throughout their lives. The cat preferred feasting on mammoths and mastodons (Yurk liked that kind of meat himself), but when confronted with her only threat, her only true enemy, the cat knew she was wise to be ready. Life of any sort was difficult enough—not being alert and wary at all times was certainly an invitation to disaster—for both of them.

It was so warm.

But now things were different. Neither the cat nor Yurk was attentive to each other; they were not as alert, as wary of each other as they had been in the past. Each knew, in their own way, that the days of hunting and protecting themselves were behind them—food certainly was not a consideration with either. No, food was not a problem; they were not hungry. Afraid? No, not really.

The cat continued her climb and finally reached the summit. She stood looking out upon the swamp (with one eye semifocused on Yurk). Yurk stood below, looking out on the swamp (aware of her presence as well).

They both knew, in their own way that things were different. Hell, they could feel the difference; it was so warm.

Warm—yes, it was warm. For their entire lives, they had never known such warmth, had never seen the snow and ice melt, had never witnessed the swampy landscape now before them. Their world was different—fearfully and wonderfully changed.

The warming trend had actually begun about 2 or 3 years earlier, though Yurk and the cat had barely been aware of it, because the increase in temperatures had been subtle—just about a half degree Fahrenheit each 3 months or so. But now the difference was obvious. The temperature was a least 10 degrees warmer than they had ever experienced—thus the melt, the freshly uncovered swamp, the rock-strewn ledge, and the warmth, of course.

The cat and Yurk stood for a time, gazing out at the swamp. What this change would mean to their clan and mates—those to follow—they were not capable of determining. What this change would bring to their world, they were not capable of speculating. So they stood, until Yurk sat back down on the ground, his back against the deadfall, and the cat lay down on the heated rocks of the ledge; they were both exhausted, tired, worn out—old, so old—and warm—too warm.

About an hour later, as darkness fell total upon the blackened, spectral landscape before them, they both went to sleep—the sleep of the dead—and their own warmth turned cold.

The ambient temperature continued to rise, even now that it was dark, night. A night that when ended would bring the dawn of a new day—and the dawn of a new era.

It was so warm and getting warmer.

Spellman (1999)

INTRODUCTION

Are we headed for warmer times or colder times? Is global climate change actually happening, and if so, do we really need to worry about it? Are the tides rising? Does the ozone hole portend disaster right around the corner?

These days, many people are beginning to ask a variety of questions related to climate change. Such questions seem reasonable when you consider the constant barrage of newspaper headlines, magazine articles, and television news reports we have been exposed to in recent years. Recently, for example, El Nino (and its devastation of the West Coast of the United States and Peru and Ecuador) has received a lot of media and scientific attention. On the other side of the coin, it has helped reduce the usual number, magnitude, and devastation of hurricanes that annually blast the East Coast of the United States—though the people affected by the ice storms in upstate New York and Canada and the tornado victims in Florida in the winter of 1998 probably did not consider themselves very lucky.

What is going on? What does all this mean? We have plenty of theories and doomsdayers out there, but are they correct? If they are correct, what does it all mean? Does anyone really know the answers? Is there anything we can do?

Not really.

Should we be concerned?

Yes.

Should we panic?

No.

Should we take decisive action? Should we practice feel-good science instead of good science?

Not exactly.

What can we do? Is there anything we can do?

Yes. We can study the facts, the issues, the possible consequences—we can do all this, but we need to let scientific fact, common sense, and cool-headedness prevail. Shooting from the hip is not called for here, makes little sense, and could have *Titanic* consequences for us all.

One thing is certain: We cannot abandon ship. The SS Earth is the only ship around and whether we like it or appreciate it does not really matter, we are all passengers headed in the same direction. Hopefully our journey will continue in such a manner that we and SS Earth can continue to make headway with fair winds and following seas.

Will we be able to accomplish this? Maybe a better question: Is there anything we can do about it?

The answer, of course, is yes. We can do something—and we will because what other choice do we have?

The only question that really has any merit here is: Will we take the correct action before it is too late? The key words are *correct action*.

TABLE 12.1
Geologic Eras and Periods

Era	Period	Millions of Years before Present
Cenozoic	Quaternary	2.5–present
	Tertiary	65–2.5
Mesozoic	Cretaceous	135–65
	Jurassic	190–135
	Triassic	225–190
Paleozoic	Permian	280–225
	Pennsylvanian	320–280
	Mississippian	345–320
	Devonian	400–345
	Silurian	440–400
	Ordovician	500–440
	Cambrian	570–500
	Precambrian	4,600–570

TABLE 12.2
Geological Epochs

Epochs	Million Years Ago
Holocene	.01–0
Pleistocene	1.6–.01
Pliocene	5–1.6
Miocene	24–5
Oligocene	35–24
Eocene	58–35
Paleocene	65–58

In this chapter, global climate change related to our atmosphere and its problems, actual and potential, are discussed. Consider this: any damage we do to our atmosphere affects the other three mediums—water, soil, and biota. Thus, the endangered atmosphere (if it is endangered) is a major concern to all of us.

THE PAST

Before we begin our discussion of the past, we need to define the era we refer to when we say "the past." Table 12.1 gives the entire expanse of time from earth's beginning to present. Table 12.2 provides the sequence of geological epochs over the past 65 million years, as dated by modern methods. The Paleocene through Pliocene together make up the Tertiary period; the Pleistocene and the Holocene compose the Quaternary period.

When most people think about climatic conditions in the ancient past, they generally think of two eras: the Ice Age and the period of the dinosaurs. Of course, those two ages take up only a tiny fraction of the time earth has been spinning around the sun. Let us look at what we know about the past. In the first place, geological history has shown periods when normal climate of the earth was so warm that subtropical weather reached to 60° north and south latitude, and there was a total absence of polar ice.

Most people are unaware that it is only during less than about 1% of the earth's history, glaciers have advanced and reached as far south as what is now the temperate zone of the northern hemisphere. The latest such advance (which started about 1,000,000 years ago) was marked by geological upheaval and (arguably) the beginning of humans. During this time, vast ice sheets advanced and retreated over the continents. The following takes a closer look at these ice ages.

A TIME OF ICE

The oldest known glacial epoch occurred nearly 2 billion years ago. In southern Canada, extending east to west about 1,000 miles is a series of deposits of glacial origin. Within the last billion years or so, the earth has experienced at least six major phases of massive, significant climatic cooling and subsequent glaciation, which apparently occurred at intervals of about 150 million years. Each may have lasted as long as 50 million years.

In more recent times (Pleistocene epoch to present), examination of land and oceanic sediment core samples clearly indicate that numerous alterations between warmer and colder conditions have occurred over the last 2 million years (during middle and early Pleistocene epoch). At least eight such cycles have occurred in the last million years, with the warm part of the cycle lasting only a relatively short time.

During the Pleistocene epoch (what we commonly call the Great Ice Age), a series of ice advances began, at times covering over one-fourth of the earth's land surface with great sheets of ice thousands of feet thick. Glaciers moved across North America many times, reaching as far south as the Great Lakes, and an ice sheet thousands of feet thick spread over Northern Europe, sculpting the land and leaving behind its giant footprints in the form of numerous lakes and swamps and its toe prints in the form of terminal moraines as far south as Switzerland. Evidence appears to indicate that each succeeding glacial advance was more severe than the previous one. The most severe began about 50,000 years ago and ended about 10,000 years ago. Several glacial advances were separated by interglacial stages, during which the ice melted and temperatures were, on average, higher than today.

Temperatures were higher than today? Yes. Keep this important point in mind as we proceed.

Although scientists consider the earth still to be in a glacial stage (because one-tenth of the globe's surface is still covered by glacial ice), ever since the climax of the last glacial advance, the ice sheet has been in a stage of retreating, and world climates, although fluctuating, are slowing warming.

How do we know the ice sheet is in a stage of retreating? We know from our observations and well-kept records that clearly show that the last hundred years have seen marked worldwide retreat of ice. Swiss resorts built during the early 1900s to offer scenic views of glaciers now have no ice in sight. Glacier National Park in Montana, world famous for its 50 glaciers and 200 lakes, is not quite the same place it was a hundred years ago. In 1937, a 10-foot pole was driven into the ground at the terminal edge of one of the main glaciers. Today this pole is still in place, but the glacier has retreated several hundred feet back up the slope of the mountain. If this glacial retreat continues and all the ice melts, sea levels would rise more than 200 feet, flooding many of the world's major cities. New York and Boston would then become aquariums.

What causes an ice age? The cause of these periodic ice ages is a deep enigma of earth history. Scientists have advanced many theories ranging from changing ocean currents to sunspot cycles. One fact that is absolutely certain, however, is that an ice age event occurs because of a change in earth's climate. But what brings about such a drastic change? To answer this question, we need to take a closer look at some factors related to climate and climate change.

Climate results from the uneven distribution of heating over the surface of the earth caused by the earth's tilt. This tilt is the angle between the earth's rotational axis and its orbital plane around the sun. Currently this angle is 23.5 degrees.

Has this angle always been 23.5 degrees? No. That is the point of our discussion here—the angle has changed (we will discuss this in greater detail shortly).

Let us get back to climate and climate change.

Long-term climate is also affected by the heat balance of the earth, which is driven mostly by the concentration of carbon dioxide (CO_2) in the atmosphere. Climate change can result if the pattern of solar radiation is changed or if the amount of CO_2 changes. Abundant evidence that the earth does undergo climatic change exists. Climatic change can be a limiting factor for the evolution of many species.

As stated previously, there is abundant evidence (primarily from soil core samples and topographical formations) that the earth undergoes climatic change. Climate change includes events such as periodic ice ages, characterized by glacial and interglacial periods. Major glacial periods lasted up to 100,000 years, with a temperature decrease of about 9°F, and most of the planet was covered with ice. Minor periods lasted up to 12,000 years, with about a 5°F decrease in temperature, and ice covered 40° latitude and above. Smaller periods such as the Little Ice Age occurred from about 1000 to 1850 AD, when there was about a 3.8°F drop in temperature. Despite its name, Little Ice Age was not a true glacial period but rather a time of severe winters and violent storms.

We are presently in an interglacial stage that may be reaching its apogee. The earth has gone through a series of glacial periods, and from the best information available to us, these periods are cyclical. What does that mean in the long run? No one knows for sure, but let us look at the effects of ice ages.

Ice ages bring about changes in sea levels. A full-blown ice age can change sea level by about 100 meters, which would expose the continental shelves. The exposed continental shelves' composition would be changed because of increased deposition during melt. The hydrological cycle would change because less evaporation would occur. Significant changes in landscape would occur, such as the creation of huge formations on the scale of the Great Lakes. Drainage patterns throughout most of the world would change, with possible massive flooding episodes. Topsoil characteristics would change—glaciers deposit rock and grind away soil.

Are these changes significant? Many areas would be devastatingly affected by the changes that an ice age could bring, in particular, Northern Europe, Canada, Seattle, around the Great Lakes, and near coastal regions.

Let us go back to the main question: What causes ice ages? The answer: We are not sure, but there are some theories. Scientists point out, for example, in order to generate a full-blown ice age (a massive ice sheet covering most of the globe), certain periodic or cyclic events or happenings would have to take place. The periodic fluctuations referred to would have to affect the solar cycle, for instance. However, we have no definitive evidence that this has ever occurred.

Another theory speculates that periods of widespread volcanic activity generate masses of volcanic dust that would block or filter heat from the sun, and thus cool down the earth.

Some speculate that the carbon dioxide cycle would have to be periodic/cyclic to bring about periods of climate change. References to a so-called factor 2 reduction, causing a 7°F temperature drop worldwide, have been made.

Others speculate that another global ice age could be brought about by increased precipitation at the poles, caused by changing orientation of continental land masses.

Others theorize that a global ice age would result if there were changes in mean temperatures of ocean currents. But the question is how? By what mechanism?

So what are the most probable causes of ice ages on earth? According to the *Milankovitch hypothesis*, the occurrence of an ice age is governed by a combination of factors: (1) the earth's change of altitude in relation to the sun—the way it tilts in a 41,000-year cycle and at the same time wobbles on its axis in a 22,000-year cycle, making the time of its closest approach to the sun come at different seasons; and (2) the 92,000-year cycle of eccentricity in its orbit around the sun, changing it from an elliptical to a near circular orbit, with the severest period of an ice age coinciding with the approach to circularity.

So what does all this mean? What this means is we have a lot of speculation about ice ages and their causes and effects. We do not have any speculation about the fact that they actually occurred and that they caused things to occur (formation of the Great Lakes, etc.), but there is a lot we do not know—the old "we don't know what we don't know" paradox.

Many possibilities exist. At this point, no single theory is sound, and doubtless, many factors are involved. But we should keep in mind that we are possibly still in the Pleistocene Ice Age. It may reach another maximum in another 60,000+ years or so. This issue will be revisited in the next section.

WARM WINTER

Maybe you have seen the headlines: "1997 Was the Warmest Year on Record," "Scientists Discover Ozone Hole Is Larger Than Ever," "Record Quantities of Carbon Dioxide Detected in Atmosphere," or "January 1998 Was the Third Warmest January on Record." Have you seen other reports stating that research indicates that we are undergoing a cooling trend?

In the previous section, we discussed several possible causes of glaciation and subsequent climatic cooling. In most scenarios discussed, we were left with the old paradox: We don't know what we don't know. Now it is time to discuss how we know what we think we know about climatic change.

WHAT WE THINK WE KNOW ABOUT GLOBAL CLIMATE CHANGE

Two large-scale environmentally significant events took place in 1997: the return of El Nino and the Kyoto Conference: Summit on Global Warming and Climate Change. As to El Nino, 1997 and 1998 news reports have blamed this phenomenon for just about everything and anything that has to do with weather conditions throughout the world. Some of these occurrences are indeed El Nino related or generated: the out-of-control fires, droughts, floods and the stretches of dead coral, the lack of fish in the water, and few birds around certain Pacific atolls. Few would argue that the devastating storms that struck the west coasts of South America, Mexico, and California were not El Nino related. Additionally, few argue against El Nino's affect on the 1997 hurricane season, one of the mildest on record. However, other anomalies or occurrences (such as lower or higher plant growth in certain regions of the globe and other absurdities, like the appearance of a double rainbow in certain areas) blamed on El Nino certainly are suspect, if not totally ridiculous.

On December 7, 1997, the Associated Press reported that while delegates at the global climate conference in Kyoto haggled over greenhouse gases and emission limits, a compelling—and so far unanswered—question has emerged: Is global warming fueling El Nino?

Nobody is really sure because we need more information, more data than we have today. One thing seems certain (based on our paltry amount of recorded data): El Nino is getting stronger and more frequent.

Some scientists fear that the increasing frequency and intensity of El Nino (based on records showing that two of this century's three worst El Ninos have come recently, in 1982 and 1997) may be linked to global warming. Experts at the Kyoto Conference say the hotter atmosphere is heating up the world's oceans, which could set the stage for more frequent and extreme El Ninos.

We have little doubt that weather-related phenomena seem to be intensifying throughout the globe. Can we be sure that this is related to global warming yet? No. The jury is still out. We need more data, more science, more time.

Is there cause for concern? Yes. According to the Associated Press coverage of the Kyoto Conference, scientist Richard Fairbanks reported that he found some starting evidence of our need for concern. During 2 months of scientific experiments on Christmas Island (the world's largest atoll in the Pacific Ocean) conducted in autumn 1997, he discovered a frightening scene. The water surrounding the atoll was 7 degrees (F) higher than average for that time of year, throwing the environmental system out of balance. According to Fairbanks, 40% of the coral was dead, the warmer water had killed off or driven away fish, and the atoll's normally plentiful bird population was almost completely gone.

Few would argue with the impact that El Nino is having on the globe. However, we are not certain that it is caused or intensified because of global warming.

The natural question now shifts to: What do we know about global warming and climate change?

USA Today (December 1997) reported on the results of a report issued by the Intergovernmental Panel on Climate Change and an interview with Jerry Mahlman of the National Oceanic and Atmospheric Administration and Princeton University, in which the following information about what most scientists agree on was obtained (p. A-2):

- There is a natural greenhouse effect (first discovered by Joseph Fourier in 1824) and scientists know how it works, and without it, earth would freeze.
- The earth undergoes normal cycles of warming and cooling on grand scales. Ice ages occur every 20,000 to 100,000 years.
- Globally, average temperatures have risen 1 degree in the past 100 years, within the range that might occur normally.

- The level of man-made carbon dioxide in the atmosphere has risen 30% since the beginning of the industrial revolution in the nineteenth century, and is still rising.
- Levels of man-made carbon dioxide will double in the atmosphere over the next 100 years and generate a rise in global average temperatures of about 3.5°F (larger than the natural swings in temperature that have occurred over the past 10,000 years).
- By 2050, temperatures will rise much higher in northern latitudes than the increase in global average temperatures. Substantial amounts of northern sea ice will melt, and snow and rain in the northern hemisphere will increase.
- As the climate warms, the rate of evaporation will rise, further increasing warming. Water vapor also reflects heat back to earth.

WHAT WE THINK WE KNOW ABOUT GLOBAL WARMING

What is global warming? To answer this question we need to discuss greenhouse effect. We know that water vapor, carbon dioxide, and other atmospheric gases (greenhouse gases; see Table 12.3) help to warm the earth. Without this greenhouse effect, the earth's average temperature would be closer to zero than its actual 60 degrees. As gases are added to the atmosphere, the average temperature could increase, changing orbital climate.

GREENHOUSE EFFECT

To understand earth's greenhouse effect, here's an explanation most people (especially gardeners) are familiar with. In a garden greenhouse, the glass walls and ceilings are

---

TABLE 12.3
The Greenhouse Gases

Greenhouse Gas	% of Total Greenhouse Gases	Sources and % of Total Greenhouse Gases
Carbon dioxide	50	Energy from fossil fuels (35)
		Deforestation (10)
		Agriculture (3)
		Industry (2)
Methane	16	Energy from fossil fuels (4)
		Deforestation (4)
		Agriculture (8)
Nitrous oxide	6	Energy from fossil fuels (4)
		Agriculture (2)
Chlorofluorocarbons (CFCs)	20	Industry (20)
Ozone	8	Energy from fossil fuels (6)
		Industry (2)

Source: EPA (2005).

---

largely transparent to short-wave radiation from the sun, which is absorbed by the surfaces and objects inside the greenhouse. Once absorbed, the radiation is transformed into long-wave (infrared) radiation (heat), which is radiated back from the interior of the greenhouse. But the glass does not allow the long-wave radiation to escape, instead absorbing the warm rays. With the heat trapped inside, the interior of the greenhouse becomes much warmer than the air outside.

The earth's atmosphere allows much the same greenhouse effect to take place. The short-wave and visible radiation that reaches earth is absorbed by the surface as heat. The long heat waves are then radiated back out toward space, but the atmosphere instead absorbs many of them. This is a natural and balanced process and, indeed, is essential to life as we know it on earth. The problem comes when changes in the atmosphere radically change the amount of absorption, and therefore the amount of heat retained. Scientists, in recent decades speculate that this may have been happening as various air pollutants have caused the atmosphere to absorb more heat. This phenomenon takes place at the local level with air pollution, causing heat islands in and around urban centers.

As pointed out earlier, the main contributors to this effect are the greenhouse gases: water vapor, carbon dioxide, carbon monoxide, methane, volatile organic compounds (VOCs), nitrogen oxides, chlorofluorocarbons (CFCs), and surface ozone. These gases delay the escape of infrared radiation from the earth into space, causing a general climatic warming. Note that scientists stress that this is a natural process. Indeed, the earth would be 33°C cooler than it is presently if the normal greenhouse effect did not exist (Hansen et al., 1986).

The problem with earth's greenhouse effect is that human activities are now rapidly intensifying this natural phenomenon, which may lead to global warming. Debate, confusion, and speculation about this potential consequence are rampant. Scientists are not entirely sure whether the recently perceived worldwide warming trend is because of greenhouse gases or because of some other cause, or whether it is simply a wider variation in the normal heating and cooling trends they have been studying. If it continues unchecked, however, the process may lead to significant global warming, with profound effects. Human impact on greenhouse effect is real; it has been measured and detected. The rate at which the greenhouse effect is intensifying is now more than five times what it was during the last century (Hansen and Lebedeff, 1989).

GREENHOUSE EFFECT AND GLOBAL WARMING

Those who support the theory of global warming base their assumptions on man's altering of the earth's normal greenhouse effect, which provides necessary warmth for life. They blame human activities (burning of fossil fuels, deforestation, and use of certain aerosols and refrigerants) for the increased amounts of greenhouse gases. These gases have increased the amounts of heat trapped in the earth's atmosphere, gradually increasing the temperature of the whole globe.

Many scientists note that (based on recent or short-term observation) the last decade has been the warmest since temperature recordings began in the late nineteenth century, and that the more general rise in temperature in the last century has coincided with the industrial revolution, with its accompanying increase in the use of fossil fuels. Other evidence supports the global warming theory. For example, in the Arctic and Antarctica, places that are synonymous with ice and snow, we see evidence of receding ice and snow cover.

Taking a long-term view, scientists look at temperature variations over thousands or even millions of years. Having done this, they cannot definitively show that global warming is anything more than a short-term variation in earth's climate. They base this assumption on historical records that have shown the earth's temperature does vary widely, growing colder with ice ages and then warming again. On another side of the argument, some people point out that the 1980s saw nine of the twelve warmest temperatures ever recorded, and the earth's average surface temperature has risen approximately 0.6°C (1°F) in the last century (USEPA, 2005). At the same time, still others offer as evidence that the same decade also saw three of the coldest years: 1984, 1985, and 1986.

So what is really going on? We are not certain, but let us assume that we are indeed seeing long-term global warming. If this is the case, we must determine what is causing it. But here, we face a problem. Scientists cannot be sure of the greenhouse effect's causes. Global warming may simply be part of a much longer trend of warming since the last ice age. Though much has been learned in the past two centuries of science, little is actually known about the causes of the worldwide global cooling and warming that have sent the earth through a succession of major ice ages and smaller ones. We simply do not have the enormously long-term data to support our theories.

FACTORS INVOLVED WITH GLOBAL WARMING/COOLING

Right now, scientists are able to point to six factors that could be involved in long-term global warming and cooling:

1. Long-term global warming and cooling could result if changes in the earth's position relative to the sun occur (i.e., the earth's orbit around the sun), with higher temperatures when the two are closer together and lower when farther apart.
2. Long-term global warming and cooling could result if major catastrophes (meteor impacts or massive volcanic eruptions) that throw pollutants into the atmosphere that can block out solar radiation occur.
3. Long-term global warming and cooling could result if changes in albedo (reflectivity of earth's surface) occur. If the earth's surface were more reflective, for example, the amount of solar radiation radiated back toward space instead of absorbed would increase, lowering temperatures on earth.
4. Long-term global warming and cooling could result if the amount of radiation emitted by the sun changes.
5. Long-term global warming and cooling could result if the shape and relationship of the land and oceans change.
6. Long-term global warming and cooling could result if the composition of the atmosphere changes.

This last possibility, of course, relates directly to our present concern: Have human activities had a cumulative impact large enough to affect the total temperature and climate of earth? We are not certain, right now, but we are somewhat concerned and alert to the problem.

SO WHAT DOES THIS ALL MEAN?

If global warming is occurring, we can expect winters to be shorter, summers longer and warmer, and sea level will rise on the order of a foot or so in the next hundred years and will continue to do so for many hundreds of years.

Let us take a closer look at the situation. We have routine global temperature measurements for only about 100 years, and these are not too reliable because of changes in instruments and methods of observation.

The only conclusion to be drawn about our climate is that we do not know whether it is changing drastically. The key word is *drastically*. Geologically, we may be at the end of an ice age. Evidence indicates that, during interglacial cycles, there is a period when temperatures increase before they plunge. Are we ascending the peak temperature range? How about human impacts on climate? Have they become so marked that we cannot be sure that the natural cycle of ice ages (which has lasted for the last 5 million years or so) will continue? Or we may just be having a breathing spell of a few centuries before the next advance of the glaciers.

Know one knows for sure. Do you?

Interesting Point

Views on the effects of global climate change and subsequently on global warming are many and varied. It is interesting to note that in my undergrad/grad air pollution courses at Old Dominion University (ODU), we spend a lot of time discussing and researching global climate change. I have found (surprisingly, to me at least) that many students do not feel that global warming is a problem. Instead, they argue that global warming is better than global cooling—no argument there. Enduring warm winters is better than enduring cold winters—in my old age, this seems sensible. Because energy supplies are limited, it is better that the globe is warming than cooling simply because it will be difficult to keep us all warm with a limited, dwindling amount of available energy. Some might say that these are simple statements that have some semblance to that rare substance that is not so common: common sense.

REFERENCES AND RECOMMENDED READING

Associated Press. 1997. Does Warming Feed El Nino? *Virginian-Pilot* (Norfolk, VA), December 7, p. A-15.
Global warming: Policies and economics further complicate the issue. *USA TODAY*, December 1, 1997, pp. A-1, A-2.

Hansen, J. E., et al. 1986. Climate sensitivity to increasing greenhouse gases. In Barth, M. C., and Titus, J. G., eds., *Greenhouse effect and sea level rise: A challenge for this generation*. New York: Van Nostrand Reinhold.

Hansen, J. E., and Lebedeff, S. 1989. Greenhouse effect of chlorofluorocarbons and other trace gases. *Journal of Geophysical Research* 94:16417–16421.

Spellman, F. R. 1999. *The science of environmental pollution*. Boca Raton, FL: CRC Press.

Spellman, F. R., and Whiting, N. 2006. *Environmental science and technology: Concepts and applications*. 2nd ed. Rockville, MD: Government Institutes.

USEPA. 2005. *Basic air pollution meteorology*. www.epa.gov/apti (accessed January 11, 2008).

Part III

Air Quality

13 Air Quality

In recent years emphasis has focused on global climate change and its potential repercussions. No doubt, the potential repercussions of global climate change are significant factors that definitely warrant our attention and our concern, but we may have another concern that dwarfs and virtually cancels out our concern about greenhouse effect and global warning and so forth. Consider this: If we are unable to breathe the air we have now and in the future because of poor air quality, then what difference does it make if we have ozone holes, hotter summers, warmer winters, melting ice caps, and rising tides?

INTRODUCTION

When undertaking a comprehensive discussion of air, the discussion begins and ends with the earth's atmosphere—and to a point we have done just that. In this chapter concepts are covered that will enable us to better understand the anthropogenic impact of pollution on the atmosphere, which in turn will enable us to better understand the key parameters used to measure air quality. Obviously, having a full understanding of air quality is essential. To set the stage for information to follow in subsequent chapters related to air pollution and air pollution control, we must review a few basic concepts.

EARTH'S HEAT BALANCE

The energy expended in virtually all atmospheric processes is originally derived from the sun. This energy is transferred by radiation of heat in the form of electromagnetic waves. The radiation from the sun has its peak energy transmission in the visible wavelength range (038 to 0.78 micrometers (μm)) of the electromagnetic spectrum. However, the sun also releases considerable energy in the ultraviolet and infrared regions. Ninety-nine percent of the sun's energy is emitted in wavelengths between 0.15 to 40 μm. Furthermore, wavelengths longer than 2.5 μm are strongly absorbed by water vapor and carbon dioxide in the atmosphere. Radiation at wavelengths less than 0.29 μm is absorbed high in the atmosphere by nitrogen and oxygen. Therefore, solar radiation striking the earth generally has a wavelength between 0.29 and 2.5 μm (EPA, 2005).

Since energy from the sun is always entering the atmosphere, the earth would overheat if all this energy were stored in the earth-atmosphere system. So, energy must eventually be released back into space. On the whole, this is what happens—approximately 50% of the solar radiation entering the atmosphere reaches earth's surface, either directly or after being scattered by clouds, particulate matter, or atmospheric gases. The other 50% is either reflected directly back or absorbed in the atmosphere and its energy reradiated back into space at a later time as infrared radiation. Most of the solar energy reaching the surface is absorbed and must be returned to space to maintain *heat balance* (aka *radiational balance*). The energy produced within the earth's interior (from hot mantle area via convection and conduction), which reaches the earth's surface (about 1% of that received from the sun), must also be lost.

Reradiation of energy from the earth is accomplished by three energy transport mechanisms: radiation, conduction, and convection. *Radiation* of energy, as stated earlier, occurs through electromagnetic radiation in the infrared region of the spectrum. The crucial importance of the radiation mechanism is that it carries energy away from earth on a much longer wavelength than that which carries the solar energy (sunlight) that brings energy to the earth and, in turn, works to maintain the earth's heat balance. The earth's heat balance is of particular interest to us in this text because it is susceptible to upset by human activities.

A comparatively smaller but significant amount of heat energy is transferred to the atmosphere by conduction from the earth's surface. *Conduction* of energy occurs through the interaction of adjacent molecules with no visible motion accompanying the transfer of heat; for example, the whole length of a metal rod will become hot when one end is held in a fire. Because air is a poor heat conductor, conduction is restricted to the layer of air in direct contact with the earth's surface. The heated air is then transferred aloft by *convection*, the movement of whole masses of air, which may be either relatively warm or cold. Convection is the mechanism by which abrupt temperature variations occur when large masses of air move across an area. Air temperature tends to be greater near the surface of the earth and decreases gradually with altitude. A large amount of the earth's surface heat is transported to clouds in the atmosphere by conduction and convection before being lost ultimately by radiation, and this redistribution of heat energy plays an important role in weather and climate conditions.

The earth's average surface temperature is maintained at about 15°C because of atmospheric greenhouse effect. Greenhouse effect occurs when the gases of the lower atmosphere transmit most of the visible portion of incident sunlight in the same way as the glass of a garden greenhouse. The warmed earth emits radiation in the infrared region, which is selectively absorbed by the atmospheric gases whose absorption spectrum is similar to that of glass. This absorbed energy heats the atmosphere and helps maintain the earth's temperature. Without this greenhouse effect, the surface temperature would average around −18°C. Most of the absorption of infrared energy is performed by water molecules in the atmosphere. In addition to the key role played by water molecules, carbon dioxide, although to a lesser extent, also is essential in maintaining the heat balance. Environmentalists and others concerned with environmental issues are worried that an increase in the carbon dioxide level in the atmosphere could prevent sufficient energy loss, causing damaging increases in the earth's temperature. This phenomenon, commonly known as anthropogenic greenhouse effect (see Chapter 12), may occur from elevated levels of carbon dioxide caused by increased use of fossil fuels and the reduction in carbon dioxide absorption because of destruction of the rainforest and other forest areas.

INSOLATION (EPA, 2005)

The amount of incoming solar radiation received at a particular time and location in the earth-atmosphere system is called insolation. Insolation is governed by four factors:

- Solar constant
- Transparency of the atmosphere
- Daily sunlight duration
- Angle at which the sun's rays strike the earth

SOLAR CONSTANT

The *solar constant* is the average amount of radiation received at a point, perpendicular to the sun's rays, that is located outside the earth's atmosphere at the earth's mean distance from the sun. The average amount of solar radiation received at the outer edge of the atmosphere would vary slightly depending on the energy output of the sun and the distance of the earth relative to the sun. Due to the eccentricity of the earth's orbit around the sun, the earth is closer to the sun in January than in July. Also, the radiation emitted from the sun varies slightly, probably less than a few percent. These slight variations that affect the solar constant are trivial considering the atmospheric properties that deplete the overall amount of solar radiation reaching the earth's surface. Transparency of the atmosphere, duration of daylight, and the angle at which the sun's rays strike the earth are much more important in influencing the amount of radiation actually received, which in turn includes the weather.

TRANSPARENCY

Transparency of the atmosphere does have an important bearing upon the amount of insolation that reaches the earth's surface. The emitted radiation is depleted as it passes through the atmosphere. Different atmospheric constituents absorb or reflect energy in different ways and in varying amounts. Transparency of the atmosphere refers to how much radiation penetrates the atmosphere and reaches the earth's surface without being depleted.

Did You Know?

Some of the radiation received by the atmosphere is reflected from the tops of clouds and from the earth's surface, and some is absorbed by molecules and clouds.

As mentioned in Chapter 11, the general reflectivity of the various surfaces of the earth is referred to as the albedo. Albedo is defined as the fraction (or percentage) of incoming solar energy that is reflected back to space. Different surfaces (water, snow, sand, etc.) have different albedo values (see Table 11.1). For the earth and atmosphere as a whole, the average albedo is 30% for average conditions of cloudiness over the earth. This reflectivity is greatest in the visible range of wavelengths.

Some of the gases in the atmosphere (notably water vapor) absorb solar radiation, causing less radiation to reach the earth's surface. Water vapor, although comprising only about 3% of the atmosphere, on average absorbs about six times as much solar radiation as all other gases combined. The amount of radiation received at the earth's surface is therefore considerably less than that received outside the atmosphere as represented by the solar constant.

Did You Know?

The earth warms up when it absorbs energy and cools when it radiates energy. The earth absorbs and emits radiation at the same time. If the earth's surface absorbs more energy than it radiates, it will heat up. If the earth's surface radiates more energy than it absorbs, it will cool.

DAYLIGHT DURATION

The duration of daylight also affects the amount of insolation received: the longer the period of sunlight, the greater the total possible insolation. Daylight duration varies with latitude and the seasons. At the equator, day and night are always equal. In the polar regions, the daylight period reaches a maximum of 24 hours in summer and a minimum of 0 hours in winter.

ANGLE OF SUN'S RAYS

The angle at which the sun's rays strike the earth varies considerably as the sun "shifts" back and forth across the equator. A relatively flat surface perpendicular to an incoming vertical sun ray receives the largest amount of insolation. Therefore, areas at which the sun's rays are oblique receive less insolation because the oblique rays must pass through a thicker layer of reflecting and absorbing atmosphere and are spread over a greater surface area. This same principle also applies to the daily shift of the sun's rays. At solar noon, the intensity of insolation is greatest. In the morning and evening hours, when the sun is at a low angle, the amount of insolation is small.

HEAT DISTRIBUTION

The earth, as a whole, experiences great contrasts in heat and cold at any particular time. Warm, tropical breezes blow at the equator while ice caps are forming in the polar regions. In fact, due to the extreme temperature differences at the equator and the poles, the earth-atmosphere system resembles a giant heat engine. Heat engines depend on hot-cold contrasts to generate power. As you will see, this global heat engine influences the major atmospheric circulation patterns as warm air is transferred to cooler areas. Different parts of the earth receiving different amounts of insolation account for much of this heat imbalance. As discussed earlier, latitude, the seasons, and daylight duration cause different locations to receive varying amounts of insolation.

DIFFERENTIAL HEATING

Not only do different amounts of solar radiation reach the earth's surface, but different earth surfaces absorb heat energy at different rates. For example, land masses absorb and store heat differently than water masses. Also, different types of land surfaces vary in their ability to absorb and store heat. The color, shape, surface texture, vegetation, and presence of buildings can all influence the heating and cooling of the ground. Generally, dry surfaces heat and cool faster than moist surfaces. Plowed fields, sandy beaches, and paved roads become hotter than surrounding meadows and wooded areas. During the day, the air over a plowed field is warmer than that over a forest or swamp; during the night, the situation is reversed. The property of different surfaces that causes them to heat and cool at different rates is referred to as *differential heating.*

Absorption of heat energy from the sun is confined to a shallow layer of land surface. Consequently, land surfaces heat rapidly during the day and cool quickly at night. Water surfaces, on the other hand, heat and cool more slowly than land surfaces for the following reasons:

- Water movement distributes heat.
- The sun's rays are able to penetrate the water surface.
- More heat is required to change the temperature of water due to its higher specific heat. (It takes more energy to raise the temperature of water than it does to change the temperature of the same amount of soil.)
- Evaporation of water occurs, which is a cooling process.

TRANSPORT OF HEAT

Earlier, it was pointed out that in addition to radiation, heat is transferred by conduction, convection, and advection. These processes affect the temperature of the atmosphere near the surface of the earth. *Conduction* is the process by which heat is transferred through matter without the transfer of matter itself. For example, the handle of an iron skillet becomes hot due to the conduction of heat from the stove burner. Heat is conducted from a warmer object to a cooler one. Heat transfer occurs when matter is in motion. Air that is warmed by a heated land surface (by conduction) will rise because it is lighter than the surrounding air. This heated air rises, transferring heat vertically. Likewise, cooler air aloft will sink because it is heavier than the surrounding air. This goes hand in hand with rising air and is part of heat transfer by convection. Meteorologists also use the term *advection* to denote heat transfer that occurs mainly by horizontal motion rather than by vertical movement or air (convection).

GLOBAL DISTRIBUTION OF HEAT

As mentioned before, the world distribution of insolation is closely related to latitude. Total annual insolation is greatest at the equator and decreases toward the poles. The amount of insolation received annually at the equator is over four times that received at either of the poles. As the rays of the sun shift seasonally from one hemisphere to the other, the zone of maximum possible daily insolation moves with them. For the earth as a whole, the gains in solar energy equal the losses of energy back into space (heat balance). However, since the equatorial region does gain more heat than it loses and the poles lose more heat than they gain, something must happen to distribute heat more evenly around the earth. Otherwise, the equatorial regions would continue to heat and the poles would continue to cool. Therefore, in order to reach equilibrium, a continuous large-scale transfer of heat (from low to high altitudes) is carried out by atmospheric and oceanic circulations.

The atmosphere drives warm air poleward and brings cold air toward the equator. Heat transfer from the tropics poleward takes place throughout the year, but at a much slower rate in summer than in winter. The temperature difference between low and high latitudes is considerably smaller in summer than in winter (only about half as large in the northern hemisphere). As would be expected, the winter hemisphere has a net energy loss and the summer hemisphere a net gain. Most of the summertime gain is stored in the surface layers of land and ocean, mainly in the ocean.

The oceans also play a role in heat exchange. Warm water flows poleward along the western side of an ocean basin, and cold water flows toward the equator on the eastern side. At higher latitudes, warm water moves poleward on the eastern side of the ocean basin and cold water flows toward the equator on the western side. The oceanic currents are responsible for about 40% of the transport of energy from the equator to the poles. The remaining 60% is attributed to the movement of air.

BASIC AIR QUALITY

The quality of the air we breathe is not normally a concern to us unless we detect something unusual about the air (its odor, its taste, or it makes breathing difficult or uncomfortable) or unless we have been advised by authorities or the news media that there is cause for concern. Air pollutants in the atmosphere cause great concern because of potential adverse effects on our health. To have good air quality is a plus for any community. Good air quality attracts industry as well as people who are looking for a healthy place to live and raise a family. It is not unusual to see or hear advertisements that push a locality's "clean" or "fresh" air as being pollution-free.

Note that although most people do seek an environment that has clean or fresh air and is pollution-free to live in, this is not always the case for everyone. A good example of this exception is in the Los Angeles Basin. Before Los Angeles became the mega-city it is today, local inhabitants named the Basin area the Valley of the Smokes from the campfires and settlements. This early warning about adverse climatic conditions did not stop settlements. Today, a stranger to Los Angeles need only pull into the Basin area, step out of his or her air-conditioned super-van, place feet on terra firma (exception during a quake), and take in one huge breath of air—then let the coughing and gagging begin. Have you ever breathed in diesel fumes mixed with other fumes? If you have, you are probably in LA—welcome to the Los Angeles Basin—home of super-smog.

Because of the large numbers of people who decided to make the LA Basin their homes, Los Angeles and California have enacted probably the most restrictive air pollution requirements anywhere.

Air quality is impacted by those things we can see readily by eye (smoke, smog, etc.) those things that can only be seen under the microscope (pollen, microbes, dust, etc.), and those substances we cannot see (ozone, carbon dioxide, sulfur dioxide, etc.). These compounds are heavily regulated, and it seems with each passing day the U.S. Environmental Protection Agency or other regulatory authority poses some new regulation for a new or old compound. When you watch a local forecast on television these days, it is not unusual to hear reference to the local air quality index.

As stated previously, air pollutants in the air we breathe cause great concern because of potential adverse effects on human health. These adverse health effects include acute conditions such as respiratory difficulties and chronic effects such as emphysema and cancer. Although health concerns related to air pollution are usually at the top of any concerned person's list, we must keep in mind that air pollution has adverse impacts on other aspects of our environment that are important to us, such as on vegetation, materials, and degradation of visibility.

In any discussion of air quality, certain specific areas must be addressed. For example, any discussion about air quality that does not include a discussion of types of air quality management (regulations), air pollutants, air pollution effects on bio-diversity (life), air pollution control technology, and indoor air quality is a hollow effort. Thus, to ensure that our effort is not hollow, these topics and others will be covered in detail in the remaining chapters.

REFERENCES AND RECOMMENDED READING

Ahrens, C. D. 2006. *Meteorology today: An introduction to weather, climate, and the environment.* 8th ed. Boston, MA: Thompson, Brooks/Cole.

EPA. 2005. Basic air pollution meteorology. www.epa.gov/apti (accessed January 12, 2008).

Oke, T. R. 1992. *Boundary layer climates.* 2nd ed. New York: Routledge.

Spellman, F. R., and Whiting, N. 2006. *Environmental science and technology: Concepts and applications.* 2nd ed. Rockville, MD: Government Institutes.

14 Air Quality Management

This we know: All things are connected like the blood that unites us. We did not weave the web of life, we are merely a strand in it. Whatever we do to the web, we do to ourselves.

We love this earth as a newborn loves its mother's heartbeat. If we sell you our land, care for it as we have cared for it. Hold in your mind the memory of the land as it is when you receive it.

Preserve the land and the air and the rivers for your children's children and love it as we have loved it.

Chief Seattle, mid-1850s

INTRODUCTION

We have found that to preserve the land and the air and the rivers for our children's children and love it as Chief Seattle and his people did, we must properly manage these valuable and crucial natural resources. We have ignored the danger signs for too long, but over the last few decades we have begun the attempt to control and manage our essential resources.

Proper air quality management includes several different areas related to air pollutants and their control. For example, we can mathematically model to predict where pollutants emitted from a source will be dispersed in the atmosphere and eventually fall to the ground and at what concentration. We have found that pollution control equipment can be added to various sources to reduce the amount of pollutants before they are emitted into the air. We have found that certain phenomena, such as acid rain, the greenhouse effect, and global warming, are all indicators of adverse effects to the air and other environmental mediums, which result from the excessive amount of pollutants being released into the air. We have found that we must concern ourselves not only with ambient air quality in our local outdoor environment, but also with the issue of indoor air quality.

To accomplish air quality management, we have found that managing is one thing—and accomplishing significant change improvement is another. We need to add regulatory authority, regulations, and regulatory enforcement authority to the air quality management scheme; strictly voluntary compliance is ineffective.

We cannot maintain a quality air supply without proper management, regulation, and regulatory enforcement. This chapter presents the regulatory framework governing air quality management. It provides an overview of the environmental air quality laws and regulations used to protect human health and the environment from the potential hazards of air pollution. New legislation, reauthorizations of acts, and new National Ambient Air Quality Standards (NAAQS) have created many changes in the way both government and industry manage their business. Fortunately, for our environment and for us, they are management tools that are effective—they are working to manage air quality.

CLEAN AIR ACT (CAA)

When you look at a historical overview of air quality regulations, you might be surprised to discover that most air quality regulations are recent. For example, in the United States the first attempt at regulating air quality came about through passage of the Air Pollution Control Act of 1955 (Public Law 84-159). This act was a step forward, but that's about all; it did little more than move us toward effective legislation. Revised in 1960 and again in 1962, the act was supplanted by the Clean Air Act (CAA) of 1963 (Public Law 88-206). CAA 1963 encouraged state, local, and regional programs for air pollution control but reserved the right of federal intervention, should pollution from one state endanger the health and welfare of citizens residing in another state. In addition, CAA 1963 initiated the development of air quality criteria upon which the air quality and emissions standards of the 1970s were based.

The move toward air pollution control gained momentum in 1970 first by the creation of the Environmental Protection Agency (EPA) and second by passage of the Clean Air Act of 1970 (Public Law 91-604), for which the EPA was given responsibility for implementation. The act was important because it set primary and secondary ambient air quality standards. Primary standards (based on air quality criteria) allowed for an extra margin of safety to protect public health, while secondary standards (also based on air quality criteria) were established to protect public welfare—animals, property, plants, and materials. Further discussion of these standards can be found in chapter 15.

The Clean Air Act of 1977 (Public Law 95-95) further strengthened the existing laws and set the nation's course toward cleaning up our atmosphere.

In 1990, the president signed the Clean Air Act Amendments of 1990. Specifically, the new law:

- Encourages the use of market-based principles and other innovative approaches, like performance-based standard and emission banking and trading
- Promotes the use of clean low-sulfur coal and natural gas, as well as the use of innovative technologies to clean high-sulfur coal through the acid rain program
- Reduces enough energy waste and creates enough of a market for clean fuels derived from grain and natural gas to cut dependency on oil imports by one million barrels/day
- Promotes energy conservation through an acid rain program that gives utilities flexibility to obtain needed emission reductions through programs that encourage customers to conserve energy

Under CAA 1990 several titles are listed with specific requirements. For example,

- Title 1—Specifies provisions for attainment and maintenance of National Ambient Air Quality Standards (NAAQS).
- Title 2—Specifies provisions relating to mobile sources of pollutants.
- Title 3—Covers air toxics.
- Title 4—Covers specifications for acid rain control.
- Title 5—Addresses permits.
- Title 6—Specifies stratospheric ozone and global protection measures.
- Title 7—Discusses provisions relating to enforcement.

TITLE 1: ATTAINMENT AND MAINTENANCE OF NAAQS

The Clean Air Act of 1977 has brought about significant improvements in U.S. air quality, but the urban air pollution problems of smog (ozone), carbon monoxide (CO), and particulate matter (PM_{10}) still persist. For example, currently, over 100 million Americans live in cities that are out of attainment with the public health standards for ozone.

A new, balanced strategy for attacking the urban smog problem was needed. The Clean Air Act of 1990 created this new strategy. Under these new amendments, states are given more time to meet the air quality standard (e.g., up to 20 years for ozone in Los Angeles), but they must make steady, impressive progress in reducing emissions. Specifically, it requires the federal government to reduce emissions from (1) cars, buses, and trucks; (2) consumer products such as window-washing compounds and hair spray; and (3) ships and barges during loading and unloading of petroleum products. In addition, the federal government must develop the technical guidance that states need to control stationary sources. In urban air pollution problems of smog (ozone), carbon monoxide (CO), and particulate matter (PM_{10}), the new law clarifies how areas are designated and redesignated "attainment." The EPA is also allowed to define the boundaries of "nonattainment" of areas (geographical areas whose air quality does not meet federal air quality standards designed to protect public health). CAA 1990 also establishes provisions defining when and how the federal government can impose sanctions on areas of the country that have not met certain conditions.

For ozone specifically, the new law established nonattainment area classifications ranked according to the severity of the area's air pollution problem. These classifications are:

- Marginal
- Moderate
- Serious
- Severe
- Extreme

The EPA assigns each nonattainment area one of these categories, thus prompting varying requirements the areas must comply with in order to meet the ozone standard.

Again, nonattainment areas have to implement different control measures, depending upon their classifications. Those closest to meeting the standard, for example, are the marginal areas, which are required to conduct an inventory of their ozone-causing emissions and institute a permit program. Various control measures must be implemented by nonattainment areas with more serious air quality problems; that is, the worse the air quality, the more controls areas will have to implement.

For carbon monoxide and particulate matter, CAA 1990 also establishes similar programs for areas that do not meet the federal health standard. Areas exceeding the standards for these pollutants are divided into moderate and serious classifications. Areas that exceed the carbon monoxide standard (i.e., to the degree to which they exceed it) are required primarily to implement programs introducing oxygenated rules and enhanced emission inspection programs. Likewise, areas exceeding the particulate matter standard have to (among other requirements) implement either reasonably available control measures (RACMs) or best available control measures (BACMs).

Title 1 attainment and maintenance of NAAQS requirements have gone a long way toward improving air quality in most locations throughout the United States. However, on November 27, 1996, in an effort to upgrade NAAQS for ozone and particulate matter, USEPA amended the National Ambient Air Quality Standards. Carol Browner, EPA administrator, later signed notice of rulemaking, putting into effect the two new NAAQS for ozone and particulate matter smaller than 2.5 μm diameter ($PM_{2.5}$). These rules appear at 62 FR 38651 for particulate matter and 62 FR 38855 for ozone. They are the first update in 20 years for ozone (smog), and the first in 10 years for particulate matter (soot).

Table 14.1 lists the National Ambient Air Quality Standards including the new requirements (updated as of January 2007). Note that NAAQS is important but the standards are not enforceable by themselves. They set ambient concentration limits for the protection of human health and environment-related values. However, it is important to remember that it is a very rare case where any one source of air pollutants is responsible for the concentrations in an entire area.

TITLE 2: MOBILE SOURCES

Cars, trucks, and buses account for almost half the emissions (even though great strides have been made since the 1960s in reducing the amounts) of the ozone precursors, volatile organic carbons (VOCs) and nitrogen oxides, and up to 90% of the CO emissions in urban areas. A large portion of the emission reductions gained from motor vehicle emission controls has been offset by the rapid growth in the number of vehicles on the highways and the total miles driven.

Because of the unforeseen growth in automobile emissions in urban areas, compounded with the serious air pollution problems in many urban areas, Congress made significant changes to the motor vehicle provisions of the 1977 Clean Air Act. The Clean Air Act of 1990 established even tighter pollution standards for emissions from motor vehicles. These standards were designed to reduce tailpipe emissions of hydrocarbons, nitrogen oxides, and carbon monoxide on a phased-in basis, which began with model year 1994. Automobile manufactures are also required to reduce vehicle emissions resulting from the evaporation of gasoline during refueling.

The latest Clean Air Act (1990 with 1997 amendments for ozone and particulate matter) also requires fuel quantity to be controlled. New programs were required for

---

TABLE 14.1
National Ambient Air Quality Standards (NAAQS)

Pollutant	Standard Value	
Carbon monoxide (CO)		
8-hour average	9 ppm	10 mg/m³
1-hour average	35 ppm	40 mg/m³
Lead (Pb)		
Quarterly average	1.5 µg/m³	
Nitrogen dioxide (NO₂)		
Annual arithmetic mean	0.053 ppm	100 µg/m³
Ozone (O₃)		
1-hour average	0.12 ppm	235 µg/m³
8-hour average	0.08 ppm	157 µg/m³
Particulate matter (PM₁₀)		
Annual arithmetic mean	50 µg/m³	
24-hour average	150 µg/m³	
Particulate matter (PM₂.₅)		
Annual arithmetic mean	15 µg/m³	
24-hour average	65 µg/m³	
Sulfur dioxide (SO₂)		
Annual arithmetic mean	0.03 ppm	80 µg/m³
24-hour average	0.14 ppm	365 µg/m³

Source: USEPA (2007).

---

cleaner or reformulated gasoline initiated in 1995 for the cities (nine total) with the worst ozone problems. Other cities were given the option to buy in to the reformulated gasoline program. In addition, a clean fuel car pilot program was established in California, which required the phasing in of tighter emission limits for several thousand vehicles in model year 1996 and up to 300,000 by model year 1999. The law allows these standards to be met with any combination of vehicle technology and cleaner fuels. Note that the standards became even stricter in 2001.

TITLE 3: AIR TOXICS

Toxic air pollutants (those that are hazardous to human health or the environment—carcinogens, mutagens, and reproductive toxins) were not specifically covered under Clean Air Act 1977. This situation is quite surprising and alarming when you consider that information generated as a result of SARA Title III (Superfund Section 313) indicates that in the United States more than 2 billion pounds of toxic air pollutants is emitted annually.

The Clean Air Act of 1990 offered a comprehensive plan for achieving significant reductions in emissions of hazardous air pollutants from major sources. The new law improved the EPA's ability to address this problem effectively and dramatically accelerated progress in controlling major toxic air pollutants.

The 1990 law includes a list of 189 toxic air pollutants whose emissions must be reduced. The EPA was required to publish a list of source categories that emit certain levels of these pollutants. The EPA was also required to issue maximum achievable control technology (MACT) standards for each listed source category, and the law also established a chemical safety board to investigate accidental releases of extremely hazardous chemicals.

Title 4: Acid Depostion

Let us talk about the acid rain problem for a moment. Consider the following:

In the evening, when you stand on your porch and look out on your terraced lawn and that flourishing garden of perennials during a light rainfall, you probably feel a sense of calm and relaxation hard to describe—but not hard to accept. It is probably the sound of raindrops against the roof of the house and porch, against the foliage and lawn, the sidewalk, the street, and that light wind through the boughs of the evergreens that soothes you. Whatever it is that makes you feel this way, rainfall is a major ingredient.

But someone knowledgeable or trained in environmental science might take another view of such a seemingly welcome and peaceful event. He or she might wonder whether the rainfall is as clean and pure as it should be. Is this actually rainfall, or is it rain carrying acids as strong as lemon juice or vinegar with it—capable of harming both living and nonliving things like trees, lakes, and man-made structures? This may seem strange to some folks who might wonder why anyone would be concerned about such an off-the-wall matter.

Maybe such a concern was unheard of before the industrial revolution, but today, the purity of rainfall is a major concern for many people, especially regarding acidity. Most rainfall is slightly acidic because of decomposing organic matter, the movement of the sea, and volcanic eruptions, but the principal factor is atmospheric carbon dioxide, which causes carbonic acid to form. *Acid rain* (pH < 5.6) (in the pollution sense) is produced by the conversion of the primary pollutants sulfur dioxide and nitrogen oxides to sulfuric acid and nitric acid, respectively. These processes are complex, depending on the physical dispersion processes and the rates of the chemical conversions.

Contrary to popular belief, acid rain is not a new phenomenon, nor does it result solely from industrial pollution. Natural processes—volcanic eruptions and forest fires, for example—produce and release acid particles into the air. The burning of forest areas to clear land in Brazil, Africa, and other countries also contributes to acid rain. However, the rise in manufacturing, which began with the industrial revolution, literally dwarfs all other contributions to the problem.

The main culprits are emissions of sulfur dioxide from the burning of fossil fuels, such as oil and coal, and nitrogen oxide, formed mostly from internal combustion engine emissions, which are readily transformed into nitrogen dioxide. These mix in the atmosphere to form sulfuric acid and nitric acid.

In dealing with atmospheric acid deposition, the earth's ecosystems are not completely defenseless; they can deal with a certain amount of acid through natural alkaline substances in soil or rocks that buffer and neutralize acid. The American

Midwest and southern England are areas with highly alkaline soil (limestone and sandstone) that provides some natural neutralization. Areas with thin soil and those laid on granite bedrock, however, have little ability to neutralize acid rain.

Scientists continue to study how living beings are damaged or killed by acid rain. This complex subject has many variables. We know from various episodes of acid rain that pollution can travel over very long distances. Lakes in Canada and New York are feeling the effects of coal burning in the Ohio Valley. For this and other reasons, the lakes of the world are where most of the scientific studies have taken place. In lakes, the smaller organisms often die off first, leaving the larger animals to starve to death. Sometimes the larger animals (fish) are killed directly; as lake water becomes more acidic, it dissolves heavy metals, leading to concentrations at toxic and often lethal levels. Have you ever wandered up to the local lakeshore and observed thousands of fish belly-up? Not a pleasant sight or smell, is it? Loss of life in lakes also disrupts the system of life on the land and air around them.

In some parts of the United States, the acidity of rainfall has fallen well below 5.6. In the northeastern United States, for example, the average pH of rainfall is 4.6, and rainfall with a pH of 4.0, which is 1,000 times more acidic than distilled water.

Despite intensive research into most aspects of acid rain, scientists still have many areas of uncertainty and disagreement. That is why the progressive, forward-thinking countries emphasize the importance of further research into acid rain. And that is why the 1990 Clean Air Act was strengthened to initiate a permanent reduction in SO_2 levels.

One of the interesting features of the 1990 Act is that it allowed utilities to trade allowance within their systems and buy or sell allowance to and from other affected sources. Each source must have sufficient allowances to cover its annual emissions. If not, the source is subject to excess emissions fees and a requirement to offset the excess emissions in the following year. The 1990 law also included specific requirements for reducing emissions of nitrogen oxides for certain boilers.

TITLE 5: PERMITS

The 1990 law also introduced an operating permit system similar to the National Pollution Discharge Elimination System (NPDES). The permit system has a twofold purpose: (1) to ensure compliance with all applicable requirements of the CAA and (2) to enhance the EPA's ability to enforce the act. Under the act, air pollution sources must develop and implement the program, and the EPA must issue permit program regulations, review each state's proposed program, and oversee the state's effort to implement any approved program. The EPA must also develop and implement a federal permit program when a state fails to adopt and implement its own program.

TITLE 6: OZONE AND GLOBAL CLIMATE PROTECTION

We have already discussed the global climate problem (Chapter 12), but let us take a look at the stratospheric ozone problem.

Ozone is formed in the stratosphere by radiation from the sun and helps to shield life on earth from some of the sun's potentially destructive ultraviolet (UV) radiation.

In the early 1970s, scientists suspected that the ozone layer was being depleted. By the 1980s, it became clear that the ozone shield was indeed thinning in some places, and at times even has a seasonal hole in it, notably over Antarctica. The exact causes and actual extent of the depletion are not yet fully known, but most scientists believe that various chemicals in the air are responsible.

Most scientists identify the family of chlorine-based compounds, most notably chlorofluorocarbons (CFCs) and chlorinated solvents (carbon tetrachloride and methyl chloroform), as the primary culprits involved in ozone depletion. In 1974, Molina and Rowland hypothesized the CFCs, containing chlorine, were responsible for ozone depletion. They pointed out that chlorine molecules are highly active and readily and continually break apart the three-atom ozone into the two-atom form of oxygen generally found close to earth, in the lower atmosphere.

According to Davis & Corwell (1991), the Interdepartmental Committee for Atmospheric Sciences (1975) estimates that a 5% reduction in ozone could result in nearly a 10% increase in cancer. This already frightening scenario was made even more frightening by 1987 when evidence showed that CFCs destroy ozone in the stratosphere above Antarctica every spring. The ozone hole had become larger, with more than half of the total ozone column wiped out and essentially all ozone disappeared from some regions of the stratosphere (Davis and Cornwell, 1991).

In 1988, Zurer reported that on a worldwide basis, the ozone layer shrunk approximately 2.5% in the preceding decade. This obvious thinning of the ozone layer, with its increased chances of skin cancer and cataracts, is also implicated in suppression of the human immune system, and damage to other animals and plants, especially aquatic life and soybean crops. The urgency of the problem spurred the 1987 signing of the Montreal Protocol by 24 countries, which required signatory countries to reduce their consumption of CFCs by 20% by 1993, and by 50% by 1998, marking a significant achievement in solving a global environmental problem.

The Clean Air Act of 1990 borrowed from EPA requirements already on the books in other regulations and mandated phaseout of the production of substances that deplete the ozone layer. Under these provisions, the EPA was required to list all regulated substances along with their ozone depletion potential, atmospheric lifetime, and global warming potentials.

TITLE 7: ENFORCEMENT

A broad array of authorities is contained within the Clean Air Act to make the law more readily enforceable. The EPA was given new authority to issue administrative penalties with fines, and field citations (with fines) for smaller infractions. In addition, sources must certify their compliance, and the EPA has authority to issue administrative subpoenas for compliance data.

CLEAN AIR ACT AMENDMENTS (EPA, 2005)

The Clean Air Act amendments require that State Implementation Plans (SIPs) be developed, the impact upon the atmosphere be evaluated for new sources, and air quality modeling analyses be performed. These regulatory programs required knowledge of the air quality in the region around a source, air quality modeling procedures, and the fate and transport of pollutants in the atmosphere. Implicit in air pollution programs is knowledge of the climatology of the area in question.

STATE IMPLEMENTATION PLANS (SIPS)

State Implementation Plans (SIPs) are federally approved plans developed by state (or local) air quality management authorities to attain and maintain the National Ambient Air Quality Standards (NAAQS). Generally, these SIPs are a state's (local) air quality rules and regulations that are considered an acceptable control strategy once approved by the Environmental Protection Agency (EPA). The purpose of SIPs is to control the amount and types of pollution for any given region of the United States.

In these types of control strategies, emission limits should be based on ambient pollutant concentration estimates for the averaging time that results in the most stringent control requirements. In all cases these concentrations estimates are assumed to be the sum of the pollutant concentrations contributed by the source and an appropriate background concentration. An air quality model is used to determine which averaging time (e.g., annual, 24 hour, 8 hour, 3 hour, 1 hour) results in the highest ambient impact. For example, if the annual average air quality standard is approached by a greater degree (percentage) than standards for other averaging times, the annual average is considered the restrictive standard. In this case, the sum of the highest estimated annual average concentration and the annual average background concentration provides the concentration that should be used to specify emission limits. However, if a short-term standard is approached by a greater degree and is thus identified as the restrictive standard, other considerations are required because the frequency of occurrence must also be taken into account.

NEW SOURCE REVIEW

New major stationary sources or major modifications to existing sources of air pollution are required by the Clean Air Act to obtain an air quality permit before construction is started. This process is called New Source Review (NSR), and it is required for any new major stationary source or major modification to an existing source regardless of whether the National Ambient Air Quality Standards (NAAQS) are exceeded. Sources located in areas that exceed the NAAQS (nonattainment areas) would undergo nonattainment New Source Review. New Source Review for major sources in areas where the NAAQS are not violated (attainment areas) would involve the preparation of a Prevention of Significant Deterioration (PSD) permit. Some sources will have the potential to emit pollutants for which their area is in attainment (or unclassifiable) as well as the potential to emit pollutants for which their area is in nonattainment. When this is the case, the source's permit will contain terms and conditions to meet both the PSD and NSR requirements because these are pollutant specific.

In most cases, any new source must obtain a nonattainment NSR permit if it will emit, or has the potential to emit, 100 tons per year or more of any regulated NSR pollutant for which that area is in nonattainment, from marginal to extreme. In areas where air quality problems are more severe, EPA has established lower thresholds for three criteria pollutants: ozone (VOCs), particulate matter (PM$_{10}$), and carbon monoxide. The significance levels are lower for modifications to existing sources.

In general, a new source located in an attainment or unclassifiable area must get a PSD permit if it will emit, or has the potential to emit, 250 tons per year (tpy) or more of any criteria or NSR regulated pollutant. If the source is on EPA's list of 28 PSD source categories, a PSD permit is required if it will or may emit 100 tpy or more of any NSR regulated pollutant. The significant levels are lower for modifications to existing sources. In addition, a PSD review would be triggered, with respect to a particular pollutant, if a new source or major modification is constructed within 10 kilometers of a class I area (see below) and would have an impact on such area equal to or greater than 1 mg/m^3 (24-hour average) for the pollutant, even through the emissions of such pollutant would not otherwise be considered significant.

Did You Know?

Ozone is not emitted by industrial sources; it is formed by volatile organic compounds (VOCs) and nitrogen oxides (known as ozone precursors) in the presence of heat and sunlight. VOCs emissions are regulated as a surrogate for ozone.

Some new sources or modifications to sources that are in attainment areas may be required to perform an air quality modeling analysis. This *air quality impact analysis* should determine if the source will cause a violation of the NAAQS or cause air quality deterioration that is greater than the available PSD increments. PSD requirements provide an area classification system based on land use for areas within the United States. These three areas are class I, class II, and class III, and each class has an established set of increments that cannot be exceeded. Class I areas consist of national parks and wilderness areas that are only allowed a small amount of air quality deterioration. Due to the pristine nature of these areas, the most stringent limits on air pollution are enforced in the class I areas. Class II areas consist of normal, well-managed industrial development. Moderate levels of air quality deterioration are permitted in these regions. Class III areas allow the largest amount of air quality deterioration to occur. When a PSD analysis is performed, the PSD increments set forth a maximum allowable increase in pollutant concentrations, which limits the allowable amount of air quality deterioration in an area. This in turn limits the amount of pollution that enters the atmosphere for a given region. In order to determine if a source of sulfur dioxide, for example, will cause an air quality violation, the air quality analysis uses the highest estimated concentration for annual averaging periods, and the second highest estimated concentration for averaging periods of 24 hours or less. The new NAAQS for PM and ozone contain specific procedures for determining modeled air quality violations.

For reviews of new or modified sources, the air quality impact analysis should generally be limited to the area where the source's impact is significant, as defined by regulations. In addition, due to the uncertainties in making concentration estimates of large downwind distances, the air quality impact analysis should generally be limited to a downwind distance of 50 kilometers, unless adverse impacts in a class I area may occur at greater distances.

AIR QUALITY MONITORING

As mentioned in the previous two sections, air quality modeling is necessary to ensure that a source is in compliance with the SIP and New Source Review requirements. When air quality modeling is required, the selection of a model is dependent on the source characteristics, pollutants emitted, terrain, and meteorological parameters. The EPA has compiled the *Guideline on Air Quality Modeling* (40 CFR 1 Appendix W), which summarizes the available models, techniques, and guidance in conducting air quality modeling analyses used in regulatory programs. This document was written to promote consistency among modelers so that all air quality modeling activities would be based on the same procedures and recommendations.

When air quality modeling is required, the specific model used (from a simple screening tool to a refined analysis) will need meteorological data. The data can vary from a few factors, such as average wind speed and Pasquill-Gifford stability categories of a mathematical representation of turbulence. Whatever model is chosen to estimate air quality, the meteorological data must match the quality of the model used. For example, average wind speed used in a simple screening model will not be sufficient for a complex refined model. An air quality modeling analysis incorporates the evaluation of terrain, building dimensions, ambient monitoring data, relevant emissions from nearby sources, and the aforementioned meteorological data.

For a dispersion model to provide useful and valid results, the meteorological data used in the model must be representative of the transport, and dispersions used in the model must be representative of the transport and dispersions characteristics in the vicinity of the source that the model is trying to simulate. The representativeness of the meteorological data is dependent on the following:

- The proximity of the meteorological monitoring site to the area under consideration
- The complexity of the terrain in the area
- The exposure of the meteorological monitoring site
- The period of time during which the data are collected

In addition, the representativeness of the data can be adversely affected by large distances between the source and the receptor of interest. Similarly, valley/mountain, land/water, and urban/rural characteristics affect the accuracy of the meteorological data for the source under consideration.

For control strategy evaluations and New Source Review, the minimum meteorological data required to describe transport and dispersion of air pollutants in the atmosphere are wind direction, wind speed, mixing height, and atmospheric stability (or related indicators of atmospheric turbulence and mixing). Because of the question of representativeness of meteorological data, site-specific data are preferable to data collected off-site. Typically 1 year of on-site data is required. If an off-site database is used (from a nearby airport, for example), 5 years of data are normally required. With 5 years of data, the model can incorporate most of the possible variations in the meteorological conditions at the site.

VISIBILITY

Visibility is the distance an observer can see along the horizon. The scattering and absorption of light by air pollutants in the atmosphere impairs visibility.

There are generally two types of air pollution that impair visibility. The first type consists of smoke, dust, or gaseous plumes that obscure the sky or horizon and are emitted from a single source or small group of sources. The second type is a widespread area of haze that impairs visibility in every direction over a large area and originates from a multitude of sources. Regardless of the type of air pollution that impairs the visibility at a particular location, any change in the meteorology or source emissions that would increase the pollutant concentration in the atmosphere will result in increased visibility impairment.

PSD class I areas have the most stringent PSD increments, and therefore must be protected not only from high pollutant concentrations, but also from the additional problems pollutants in the atmosphere can cause. Under the Clean Air Act, PSD class I areas must be evaluated for visibility impairment. This may involve a visibility impairment analysis. According to EPA regulations, visibility impairment is defined as any humanly perceptible change in visibility (visual range, contrast, or coloration) from natural conditions. Therefore, any location is susceptible to visibility impairment due to air pollution sources. Since PSD class I areas (national parks and wilderness areas) are known for their aesthetic quality, any change or alteration in the visibility of the area must be analyzed.

POLLUTANT DISPERSION

Pollutant dispersion is the process of pollutants being removed from the atmosphere and deposited onto the surface of the earth. Stack plumes contain gases and a small amount of particles that are not removed from the gas stream. When the plume emerges from the stack, these particles are carried with it. Once airborne, the particles begin to settle out and become deposited on the ground and on surface objects. There are basically two ways the particles can be deposited: dry deposition (gravitational settling) or wet deposition (precipitation scavenging). Depending on the meteorological conditions during the time of pollutant emission, these may:

1. Settle out quickly due to their weight and the effect of gravity
2. Be transported further downwind of the source due to buoyancy and wind conditions
3. Be washed out of the atmosphere by precipitation or clouds (wet deposition)

In any case, the deposition of these pollution particles is important to understand and quantify since pollutants deposited upon the ground can impact human health, vegetation, and wildlife.

Pollutant deposition concentrations must be predicted in order to minimize the risk to human health. In order to quantify the amount of pollutant deposition that occurs from stack emissions, air quality models can be used. These models determine pollution deposition based on the chemical reactivity and solubility of various gases and by using detailed data on precipitation for the areas in question.

VAPOR PLUME–INDUCED ICING

Vapor plumes are emitted from cooling towers and stacks and consist mainly of water vapor. Although pollutant concentrations are not a major concern with vapor plumes, other problems arise when vapor plume sources are located close to frequently traveled roads and populated areas. Vapor emitted from a stack is warm and misty. When meteorological conditions are favorable, the moisture in the vapor plume condenses out and settles on cooler objects (e.g., road surfaces). This phenomenon is similar to the moisture that collects on the sides of a glass of water on a warm day. If temperatures are at or below freezing when the moisture condenses, road surfaces can freeze rapidly, creating hazardous driving conditions. In addition, light winds can cause the plume to remain stagnant, creating a form of ground fog that can cause low visibility as well. Water vapor plumes that lower visibility can create hazards for aircraft, especially during critical phases of flight, including landings and takeoffs.

ATMOSPHERIC CONTAMINATION AND EMERGENCY RESPONSE

Consider the following:

Day rose heavy and hot, but the wind whispered in the field beyond the sod house as if murmuring delightful secrets to itself. A light breeze entered open windows and gently touched those asleep inside. A finger of warmth, laden with the rich, sweet odor of earth, lightly touched Juju's cheek—rousing her this morning as it had often in her 9 years of life. On most days, Juju would lay on her straw mat and daydream, languishing in the glory of waking to another day on Mother Earth. But nothing was normal on this morning. This day was different—full of surprises and excitement. Juju and her mother, Lanruh, were setting out on adventure today—and Juju could not wait.

As she stood at the foot of her makeshift bed, Juju swiftly tucked the folds of thin fabric around her slender waist and let the fall of cloth hang to her feet. She pulled her straight black hair tight in a knot at the back of her neck before she draped the end of the sari over her head.

While Juju dressed, Lanruh performed the same ritual in her small room, next to Juju's. Lanruh was excited about the day's events, too—She knew Juju was thrilled and was delighted in her daughter's pleasure and excitement. Lanruh chuckled to herself as she remembered the many times over the last few years that Juju had begged to be included, to be taken to the grand market in town. Lanruh understood Juju's excitement. Going into the town, taking it all in—the market thrilled Lanruh, too.

As they stepped out of the sod house and onto the dirt road, the scented breeze that had touched Juju's cheek earlier greeted them. They walked together, hand in hand toward town, 3 kilometers to the south.

Juju bubbled with anticipation, but she held it in, presenting the calm, serene face expected of her. Even so, every nerve in her young body reverberated with excitement.

As they walked along the road, Juju, fascinated by everything she saw, took in everything they passed in this extension of her small world. People and cattle everywhere—she had never seen so many of either! Her world had grown, suddenly—and it felt good to be alive.

As they neared town, Juju could see tall buildings. How big and imposing they were—and so many of them! In town, in places they passed, some of the streets were actually paved. Juju had never seen paved streets. This trip to town was her first city experience, and she was enthralled by all the strange and wonderful sights. As they walked along the street leading to the marketplace, Juju was overawed by the tall buildings and warehouses. "What could they all be used for?" she wondered. Some of them had sign boards above their doors, but little good that did for Juju—she could not read.

The light, following breeze had escorted Juju and Lanruh since they left home, and it was still with them as they turned toward the market. Juju could see the entrance, and the throngs of bustling people ahead, and her eyes snapped with excitement.

Suddenly, with one breath of that sweet air (was it the same sweet air that had touched her into waking only 2 hours earlier?) Juju began coughing. She clutched her throat with both hands, falling to her knees in sudden agony. Her mother was also fallen, gasping for air. The breeze that had begun her day now ended it—delivering an agent of death. But Juju did not have time to realize what was happening. She could not breath. She could not do anything—except die—and she did.

Juju, Lanruh, and over 2,000 others died within a very few minutes.

Those who died that December 3, 1984, day never knew what killed them. The several hundred others who died soon after did not know what killed them, either.

The several thousand inhabitants who lived near the marketplace, near the industrial complex, near the pesticide factory, near the chemical spill, near the release point of that deadly toxin knew little, if any of this. They knew only death and killing sickness that sorry day.

Those who survived that day were later told that a deadly chemical had killed their families, their friends, their neighbors, their acquaintances. They were killed by a chemical spill that today is infamous in the journals of hazardous materials incidents. Today, this incident is studied by everyone who has anything to do with chemical production and handling operations. We know it as Bhopal.

The dead knew nothing of the disaster—and their deaths were the result.

To prevent catastrophic incidents such as the one above, regulatory agencies require sources that have the potential to release hazardous materials into the atmosphere to implement emergency planning and response procedures. These procedures are designed to enable a facility owner to take emergency action for public protection. In addition, emergency planning can enable the facility owner to provide assessments of the emergency situation based on meteorological parameters and the airborne release of the hazardous chemicals. The emergency plans and procedures include specific meteorological measurements that must be evaluated in order to anticipate the transport and dispersion of any hazardous materials that could be emitted during an emergency situation at a site. Therefore, not only is it important to know which hazardous pollutants a facility is capable of releasing, but it is just as important to know the meteorological conditions that are prevalent at the site in order to predict how hazardous substances would be handled if accidentally released into the atmosphere. Continuous on-site meteorological data are an important factor in assessing the transport and dispersion of accidental releases. Information on wind speed and direction, atmospheric stability, and mixing height is crucial in determining the area potentially impacted by a sudden release and initiating emergency response actions such as evacuation.

REFERENCES AND RECOMMENDED READING

Davis, M. L., and Cornwell, D. A. 1991. *Introduction to environmental engineering.* New York: McGraw-Hill, Inc.

Molina, M. and Rowland, R., 1974. Stratospheric sink for chlorofluoromethanes: Chloride atom catalyzes destruction of ozone. *Nature,* 248: 810–812

Spellman, F. R., and Whiting, N. 2006. *Environmental science and technology: Concepts and applications.* Boca Raton, FL: CRC Press.

USEPA. 2005. Basic air pollution meteorology. www.epa.gov/apti (accessed January 15, 2008).

USEPA. 2007. National ambient air quality standards (NAAQS). www.epa.gov/air/criteria/html (accessed January 12, 2008).

Zurer, P. S. 1988. Studies on ozone destruction expand beyond Antarctic. *C & E News,* May, pp. 18–25.

15 Air Pollution

Thousands of chemicals are commonly used through the world for industrial, agricultural, and domestic purpose with many new ones being produced yearly. The majority of these chemicals, many of which are toxic ... eventually enter into the atmosphere and may pose a risk to the well-being of plants, animals, and microorganisms.

Barker and Tingey (1991)

INTRODUCTION

In the past, the sight of belching smokestacks was comforting to many people: more smoke equaled more business, which indicated that the economy was healthy. But many of us are now troubled by evidence that indicates that polluted air adversely affects our health. Many toxic gases and fine particles entering the air pose health hazards: cancer, genetic defects, and respiratory disease. Nitrogen and sulfur oxides, ozone, and other air pollutants from fossil fuels are inflicting damage on our forests, crops, soils, lakes, rivers, coastal waters, and buildings. Chlorofluorocarbons (CFCs) and other pollutants entering the atmosphere are depleting the earth's protective ozone layer, allowing more harmful ultraviolet radiation to reach the earth's surface. Fossil fuel combustion is increasing the amount of carbon dioxide in the atmosphere, which can have severe long-term environmental impact.

It is interesting to note that when ambient air is considered, the composition of unpolluted air is unknown to us. Humans have lived on the planet thousands of years and influenced the composition of the air through their many activities before it was possible to measure the constituents of the air.

EPA (2005) points out that in theory the air has always been polluted to some degree. Natural phenomena such as volcanoes, windstorms, the decomposition of plants and animals, and even the aerosols emitted by the ocean pollute the air. However, the pollutants we usually refer to when we talk about air pollution are those generated as a result of human activity. An *air pollutant* can be considered a substance in the air that, in high enough concentrations, produces a detrimental environmental effect. These effects can be either health effects or welfare effects. A pollutant can affect the health of humans, as well as the health of plants and animals. Pollutants can also affect nonliving materials such as paints, metals, and fabrics. An *environmental effect* is defined as a measurable or perceivable detrimental change resulting from contact with an air pollutant.

Human activities have had a detrimental effect on the makeup of air. Activities such as driving cars and trucks, burning of coal, oil, and other fossil fuels, and manufacturing chemicals have changed the composition of air by introducing many pollutants. There are hundreds of pollutants in the ambient air. Ambient air is the air to which the general public has access, i.e., any unconfined portion of the atmosphere. The two basic physical forms of air pollutants are particulate matter and gases. Particulate matter includes small solid and liquid particles such as dust, smoke, sand, pollen, mist, and fly ash. Gases include substances such as carbon monoxide (CO), sulfur dioxide (SO_2), nitrogen oxides (NO_2), and volatile organic compounds (VOCs).

Historically, many felt that the air renewed itself, through interaction with vegetation and the oceans, in sufficient quantities to make up for the influx into our atmosphere of anthropogenic pollutants. Today, however, this kind of thinking is being challenged by evidence that clearly indicates that increased use of fossil fuels, expanding industrial production, and growing use of motor vehicles are having a detrimental effect on the atmosphere, air, and environment. In this chapter, we discuss pollutant dispersal, transformation, and deposition mechanisms, and examine the types and sources of air pollutants that are related to these concerns.

ATMOSPHERIC DISPERSION, TRANSFORMATION, AND DEPOSITION (EPA, 2005)

A source of air pollution is any activity that causes pollutants to be emitted into the air. There have always been natural sources of air pollution, also known as biogenic sources. For example, volcanoes have spewed particulate matter and gases into our atmosphere for millions of years. Lightning strikes have caused forest fires, with their resulting contribution of gases and particles, for as long as storms and forests have existed. Organic matter in swamps decay and windstorms whip up dust. Trees and other vegetation contribute large amounts of pollen and spores to our atmosphere. These natural pollutants can be problematic generated pollutants or anthropogenic sources.

The quality of daily life depends on many modern conveniences. People enjoy the freedom to drive cars and travel in airplanes for business and pleasure. They expect their homes to have electricity and their water to be heated for bathing and cooking. They use a variety of products such as clothing, pharmaceuticals, and furniture made of synthetic materials. At times, they rely on services that use chemical solvents, such as the local dry cleaner and print shop. Yet the availability of these everyday conveniences comes at a price, because they all contribute to air pollution.

Air pollutants are released from both stationary and mobile sources. Scientists have gathered much information that is available on the sources, quantity, and toxicity levels of these pollutants. The measurement of air pollution is an important scientific skill, and practitioners of this skill are usually well founded in the pertinent related sciences, in modeling aspects applicable to their studies and analyses of air pollutants in the ambient atmosphere. However, to get at the very heart of air pollution, the practitioner must also be well versed in how to determine the origin of the pollutants and understand the mechanics of the pollutant dispersal, transport, and deposition.

Air pollution practitioners must constantly deal with one basic fact: air pollutants rarely stay at their release location. Instead, wind flow conditions and turbulence, local topographic features, and other physical conditions work to disperse these pollutants. So, along with having a thorough knowledge and understanding of the pollutants in question, the air pollution practitioner has a definite need for detailed knowledge of the atmospheric processes that govern their subsequent dispersal and fate. Chapter 10 has already discussed wind and its formation, a critical factor in air pollutant dispersion. In this section, we discuss several other important factors related to pollutant dispersal and fate.

Conversion of precursor substances to secondary pollutants such as ozone is an example of chemical transformation in the atmosphere. Transformations are both physical and chemical and affect the ultimate impact of originally emitted air pollutants.

Pollutants emitted to the atmosphere do not remain there forever. Two common deposition (depletion) mechanisms are *dry deposition* (the removal of both particles and gases as they come into contact with the earth's surface) and *washout* (the uptake of particles and gases by water droplets and snow and their removal from the atmosphere as precipitation that falls to the ground). Acid deposition (acid rain) is a form of pollution depletion from the atmosphere.

The following sections discuss atmospheric dispersion of air pollutants in greater detail and the main factors associated with this phenomenon, including weather, turbulence, air parcels, buoyancy factors, lapse rates, mixing, topography, inversions, plume behavior, and transport.

WEATHER

Recall that in Chapter 10, we said that air contained in earth's atmosphere is not still. Constantly in motion, air masses warmed by solar radiation rise at the equator and spread toward the colder poles where they sink, and flow downward, eventually returning to the equator. Near the earth's surface, as a result of the earth's rotation, major wind patterns develop. During the day the land warms more quickly than the sea; at night, the land cools more quickly. Local wind patterns are driven by this differential warming and cooling between the land and adjacent water bodies. Normally, onshore breezes bring cooler, denser air from over the land masses out over the waters during the night. Precipitation is also affected by wind patterns. Warm, moisture-laden air rising from the oceans is carried inland, where the air masses eventually cool, causing the moisture to fall as rain, hail, sleet, or snow.

Even though pollutant emissions may remain relatively constant, air quality varies tremendously from day to day. The determining factors have to do with weather.

Weather conditions have a significant impact on air quality and air pollution, both favorable and unfavorable, especially on local conditions. For example, on hot, sun-filled days, when the weather is calm with stagnating high-pressure cells, air quality suffers, because these conditions allow the buildup of pollutants on the ground level. When local weather conditions include cool, windy, stormy weather with turbulent low-pressure cells and cold fronts, these conditions allow the upward mixing and dispersal of air pollutants.

Weather has a direct impact on pollution levels in both mechanical and chemical ways. Mechanically, precipitation works to cleanse the air of pollutants (transferring the pollutants to rivers, streams, lakes, or the soil). Winds transport pollutants from one place to another. Winds and storms often dilute pollutants with cleaner air, making pollution levels less annoying in the area of their release. Air and its accompanying pollution (in a low-pressure cell) are also carried aloft by air heated by the sun.

When wind accompanies this rising air mass, the pollutants are diluted with fresh air. In a high-pressure cell, the opposite occurs with air and the pollutants it carries sink toward the ground. With no wind, these pollutants are trapped and concentrated near the ground where serious air pollution episodes may occur.

Chemically, weather can also affect pollution levels. Winds and turbulence mix pollutants together in a sort of giant chemical broth in the atmosphere. Energy from the sun, moisture in the clouds, and the proximity of highly reactive chemicals may cause chemical reactions, which lead to the formation of secondary pollutants. Many of these secondary pollutants may be more dangerous than the original pollutants.

TURBULENCE

In the atmosphere, the degree of turbulence (which results from wind speed and convective conditions related to the change of temperature with height above the earth's surface) is directly related to stability (a function of vertical distribution of atmospheric temperature). The stability of the atmosphere refers to the susceptibility of rising air parcels to vertical motion (attributed to high- and low-pressure systems, air lifting over terrain, or fronts and convection); consideration of atmospheric stability or instability is essential in establishing the dispersion rate of pollutants. When specifically discussing the stability of the atmosphere, we are referring to the lower boundary of the earth where air pollutants are emitted. The degree of turbulence in the atmosphere is usually classified by stability class. Ambient and adiabatic lapse rates are a measure of atmospheric stability.

Stability is divided into three classes: stable, unstable, and neutral. A *stable atmosphere* is marked by air cooler at the ground than aloft, low wind speeds, and consequently, a low degree of turbulence. A plume of pollutants released into a stable lower layer of the atmosphere can remain relatively intact for long distances. Thus, we can say that stable air discourages the dispersion and dilution of pollutants. An *unstable atmosphere* is marked by a high degree of turbulence. A plume of pollutants released into an unstable atmosphere may exhibit a characteristic looping appearance produced by turbulent eddies. A *neutrally stable atmosphere* is an intermediate class between stable and unstable conditions. A plume of pollutants released into a neutral stability condition is often characterized by a coning appearance as the edges of the plume spread out in a V shape.

The importance of the state of the atmosphere and stability's effects cannot be overstated. The ease with which pollutants can disperse vertically into the atmosphere is mainly determined by the rate of change of air temperature with height (altitude). Therefore, air stability is a primary factor in determining where pollutants will travel, and how long they will remain aloft. Stable air discourages the dispersion and dilution of pollutants. Conversely, in unstable air conditions, rapid vertical mixing takes place, encouraging pollutant dispersal, which increases air quality.

AIR PARCELS

In regards to air parcels described in this section, think of air inside a balloon as an analogy. This theoretically infinitesimal parcel is a relatively well-defined body of air (a constant number of molecules) that acts as a whole. Self-contained, it does not readily mix with the surrounding air. The exchange of heat between the parcel and its surroundings is minimal, and the temperature within the parcel is generally uniform.

BUOYANCY FACTORS

Atmospheric temperature and pressure influence the buoyancy of air parcels. Holding other conditions constant, the temperature of air (a fluid) increases as atmospheric pressure increases, and conversely decreases as pressure decreases. With respect to the atmosphere, where air pressure decreases with rising altitude, the normal temperature profile of the troposphere is one where temperature decreases with height.

An air parcel that becomes warmer than the surrounding air (for example, by heat radiating from the earth's surface), begins to expand and cool. As long as the parcel's temperature is greater than the surrounding air, the parcel is less dense than the cooler surrounding air. Therefore, it rises, or is buoyant. As the parcel rises, it expands, thereby decreasing its pressure, and therefore its temperature decreases as well. The initial cooling of an air parcel has the opposite effect. In short, warm air rises and cools, while cool air descends and warms.

The extent to which an air parcel rises or falls depends on the relationship of its temperature to that of the surrounding air. As long as the parcel's temperature is cooler, it will descend. When the temperatures of the parcel and the surrounding air are the same, the parcel will neither rise nor descend unless influenced by wind flow.

LAPSE RATES

The *lapse rate* is defined as the rate of temperature change with height. With an increase in altitude in the troposphere, the temperature of the ambient air usually decreases. On average, temperature decreases –6 to –7°C per kilometer. This is the normal lapse rate, but it varies widely depending on location and time of day. We define a temperature decrease with height as a negative elapse rate and a temperature increase with height as a positive elapse rate.

Did You Know?

How the atmosphere behaves when air is displaced vertically is a function of atmospheric stability. A stable atmosphere resists vertical motion; air that is displaced vertically in a stable atmosphere tends to return to its original position. This atmospheric characteristic determines the ability of the atmosphere to disperse pollutants emitted into it. To understand atmospheric stability and the role it plays in pollution dispersion, it is important to understand the mechanics of the atmosphere as they relate to vertical atmospheric motion.

In a dry environment, when a parcel of warm, dry air is lifted in the atmosphere, it undergoes adiabatic expansion and cooling. For the most part, a parcel of air does not exchange heat across its boundaries. Therefore, an air parcel that is warmer than the surrounding air does not transfer heat to the atmosphere. Any temperature changes that occur within the parcel are caused by increases or decreases of molecular activity within the parcel. Such changes occur adiabatically and are due only to the change in atmospheric pressure as a parcel moves vertically. The term *adiabatic* literally means impassable from, corresponding in this instance to an absence of heat transfer. That is, an adiabatic process is one in which there is no transfer of heat or mass across the boundaries of the air parcel. In an adiabatic process, compression results in heating and expansion results in cooling. A dry air parcel rising in the atmosphere cools at the dry adiabatic rate of 9.8°C/1,000 m and has a lapse rate of –9.8°C/1,000 m. Likewise, a dry air parcel sinking in the atmosphere heats up at the dry adiabatic rate of 9.8°C/1,000 m and has a lapse rate of 9.8°C/1,000 m. Air is considered dry, in this context, as long as any water in it remains in a gaseous state.

The *dry adiabatic lapse rate* is fixed, entirely independent of ambient air temperature. A parcel of dry air moving upward in the atmosphere, then, will always cool at the rate of 9.8°C/1,000 m, regardless of its initial temperature or the temperature of the surrounding air.

When the ambient lapse rate exceeds the adiabatic lapse rate, the ambient rate is said to be *superadiabatic*, and the atmosphere is highly unstable. When the two lapse rates are exactly equal, the atmosphere is said to be *neutral*. When the ambient lapse rate is less than the dry adiabatic lapse rate, the ambient lapse rate is termed *subadiabatic*, and the atmosphere is stable.

The cooling process within a rising parcel of air is assumed to be adiabatic (occurring without the addition or loss of heat). A rising parcel of air (under adiabatic conditions) behaves like a rising balloon, with the air in that distinct parcel expanding as it encounters air of lesser density until its own density is equal to that of the atmosphere that surrounds it. This process is assumed to occur with no heat exchange between the rising parcel and the ambient air (Peavy et al., 1985).

A rising parcel of dry air containing water vapor will continue to cool at the dry adiabatic lapse rate until it reaches its condensation temperature, or dew point. At this point the pressure of the water vapor equals the saturation vapor pressure of the air, and some of the water vapor begins to condense. Condensation releases latent heat in the parcel, and thus the cooling rate of the parcel slows. This new rate is called the *wet adiabatic lapse rate*. Unlike the dry adiabatic lapse rate, the wet adiabatic lapse rate is not constant but depends on temperature and pressure. In the middle troposphere, however, it is assumed to be approximately –6 to –7°C/1,000 m.

Did You Know?

The actual temperature profile of the ambient air shows the environmental lapse rate. Sometimes called the prevailing or atmospheric lapse rate, it is the result of complex interactions of meteorological factors, and is usually considered to be a decrease in temperature with height. It is particularly important to vertical motion since surrounding air temperature determines the extent to which a parcel of air rises or falls.

MIXING

Within the atmosphere, for effective pollutant dispersal to occur, turbulent mixing is important. Turbulent mixing, the result of the movement of air in the vertical dimension, is enhanced by vertical temperature differences. The steeper the temperature gradient and the larger the vertical air column in which the mixing takes place, the more vigorous the convective and turbulent mixing of the atmosphere.

TOPOGRAPHY

On a local scale, topography may affect air motion. In the United States, most large urban centers are located along sea and lake coastal areas. Contained within these large urban centers is much heavy industry. Local airflow patterns in these urban centers have a significant impact on pollution dispersion processes. Topographic features also affect local weather patterns, especially in large urban centers located near lakes, seas, and open land. Breezes from these features affect vertical mixing and pollutant dispersal. Seasonal differences in heating and cooling land and water surfaces may also precipitate the formation of inversions near the sea or lakeshore.

River valley areas are also geographical locations that routinely suffer from industry-related pollution. Many early settlements began in river valleys because of the readily available water supply and the ease of transportation afforded to settlers by river systems within such valleys. Along with settlers came industry—the type of industry that invariably produces air pollutants. These air pollutants, because of the terrain and physical configuration of the valley, are not easily removed from the valley.

Winds that move through a typical river valley are called slope winds. Slope winds, like water, flow downhill into the valley floor. At valley floor level, slope winds transform to valley winds, which flow down-valley with the flow of the river. Down-valley winds are lighter than slope winds. The valley floor becomes flooded with a large volume of air, which intensifies the surface inversion that is normally produced by radiative cooling. As the inversion deepens over the course of the night, it often reaches its maximum depth just before sunrise with the height of the inversion layer dependent on the depth of the valley and the intensity of the radiative cooling process.

Hills and mountains can also affect local airflow. These natural topographical features tend to decrease wind speed (because of their surface roughness) and form physical barriers preventing the air movement.

INVERSIONS

An inversion occurs when air temperature increases with altitude. Temperature inversions (extreme cases of atmospheric stability) create a virtual lid on the upward movement of atmospheric pollution. This situation occurs frequently but is generally confined to a relatively shallow layer. Plumes emitted into air layers that are experiencing an inversion (inverted layer) do not disperse very much as they are transported with the wind. Plumes that are emitted above or below an inverted layer do not penetrate that layer; rather, these plumes are trapped either above or below that inverted layer. High concentrations of air pollutants are often associated with inversions since they inhibit plume dispersions. Two types of inversions are important from an air quality standpoint: radiation and subsidence inversions.

Radiation inversions are the most common form of surface inversion and occur when the earth's surface cools rapidly. They prompt the formation of fog, and simultaneously trap gases and particulates, creating a concentration of pollutants. They are characteristically a nocturnal phenomenon caused by cooling of the earth's surface. On a cloudy night, the earth's radiant heat tends to be absorbed by water vapor in the atmosphere. Some of this is reradiated back to the surface. However, on clear winter nights, the surface more readily radiates energy to the atmosphere and beyond, allowing the ground to cool more rapidly. The air in contact with the cooler ground also cools, and the air just above the ground becomes cooler than the air above it, creating an inversion close to the ground, lasting for only a matter of hours. These radiation inversions usually begin to form at the worst time of the day for human concerns in large urban areas—during the late afternoon rush hour, trapping automobile exhaust at ground level and causing elevated concentrations of pollution for commuters. During evening hours, photochemical reactions cannot take place, so the biggest problem can be the accumulation of carbon monoxide. At sunrise, the sun warms the ground and the inversion begins to break up. Pollutants that have been trapped in the stable air mass are suddenly brought back to earth in a process known as *fumigation*, which can cause a short-lived, high concentration of pollution at ground level (Masters, 1991).

The second type of inversion is the *subsidence inversion*, usually associated with anticyclones (high-pressure systems); they may significantly affect the dispersion of pollutants over large regions. A subsidence inversion is caused by the characteristic sinking motion of air in a high-pressure cell. Air in the middle of a high-pressure zone descends slowly. As the air descends, it is compressed and heated. It forms a blanket of warm air over the cooler air below, thus creating an inversion (located anywhere from several hundred meters above the surface to several thousand meters) that prevents further vertical movement of air.

PLUME BEHAVIOR

One way to quickly determine the stability of the lower atmosphere is to view the shape of a smoke trail, or *plume*, from a tall stack located on flat terrain (see Figure 15.1). Visible plumes usually consist of pollutants emitted from a smokestack into the atmosphere. The formation and fate of the plume depend on a number of related factors: (1) the nature of the pollutants, (2) meteorological factors (combination of vertical air movement and horizontal airflow), (3) source obstructions, and (4) local topography, especially downwind. Overall, maximum ground-level concentrations will occur in a range from the vicinity of the smokestack to some distance downwind.

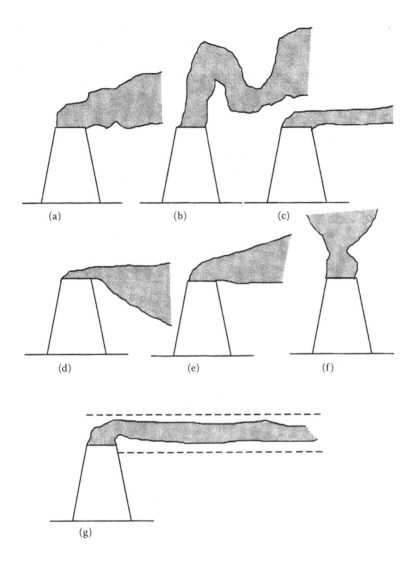

FIGURE 15.1 Seven types of classic plume behavior.

Did You Know?

One method of pollution release has received more attention than any other—pollution released from stacks. Stacks come in all sizes—from a small vent on a building's roof to a tall stack. Their function is to release pollutants high enough above the earth's surface so that emitted pollutants can sufficiently disperse in the atmosphere before reaching ground level. All else being equal, taller stacks disperse pollutants better than shorter stacks because the plume has to travel through a greater depth of the atmosphere before it reaches ground level. As the plume travels it spreads and disperses.

Figure 15.1 shows the classic types of plume behavior that are characteristic of different stability conditions. When the atmosphere is slightly stable or neutral, a typical plume "cones," as indicated in Figure 15.1a. It is likely to occur on cloudy days or on sunny days between the breakup of a radiation inversion and the development of unstable daytime conditions. When the atmosphere is highly unstable, a "looping" plume like the one shown in Figure 15.1b forms. In the looping plume, the stream of emitted pollutants undergoes rapid mixing, and the wind causes large eddies, which may carry the entire plume down to the ground, causing high concentrations close to the stack before dispersion is complete. In an extremely stable atmosphere, usually in the early morning during a radiation inversion, a "fanning" plume spreads horizontally (Figure 15.1c), with little mixing. When an inversion layer occurs a short distance above the plume source, the plume is said to be "fumigating" (Figure 15.1d). Ground-level pollutant concentrations can be very high when fumigation occurs. Sufficiently tall stacks can prevent fumigation in most cases. When inversion conditions exist below the plume source, the plume is said to be "lofting" (Figure 15.1e). When conditions are neutral, the plume issuing from a smokestack tends to rise directly into the atmosphere (Figure 15.1f). When an inversion layer prevails both above and below the plume source, the plume issuing from a smokestack tends to be "trapped" (Figure 15.1g).

Pollutants, however, rarely come from a single point source (smokestack plume). In large urban areas many plumes are generated and collectively combine into a large plume (city plume) whose dispersion represents a huge environmental challenge: the high pollutant concentrations from the city plume frequently affect human health and welfare.

Air quality problems associated with dispersion of city plumes are compounded by the presence of an already contaminated environment. Even though conventional processes normally work to disperse emissions from point sources within the city plume, because of microclimates within the city and the volume of pollutants they must handle, the conventional processes often cannot disperse effectively. Other compounding conditions present in areas where city plumes are generated (topographical barriers, surface inversions, and stagnating anticyclones) work to intensify the city plume and result in high pollutant concentrations.

PLUME RISE EQUATION

Many individuals have studied plume rise over the years. The most common plume rise formulas are those developed by Briggs, which have been extensively validated with stack plume observations (EPA, 2005). One of these that apply to buoyancy-dominated plumes is included in Equation 15.1. Plume rise formulas are to be used on plumes with temperatures greater than the ambient air temperature. The *Briggs' plume rise formula* is as follows:

$$\Delta h = \frac{1.6 F^{1/3} x^{2/3}}{u}$$ (15.1)

where
Δh = plume rise (above stack)
F = Buoyancy flux (see below; Equation 15.2)
u = average wind speed

x = downwind distance from the stack/source
g = acceleration due to gravity (9.8 m/s^2)
V = volumetric flow rate of stack gas
T_s = temperature of stack gas
T_a = temperature of ambient air

$$\text{Buoyancy flux} = F = g/\pi\ V\ (T_s - T_a/T_s) \tag{15.2}$$

TRANSPORT

Those people living east of the Mississippi River would be surprised to find out that they are breathing air contaminated by pollutants from various sources many miles from their location. Most people view pollution under the old cliché "out of sight out of mind." As far as they are concerned, if they do not see it, it does not exist. For example, assume that a person on a farm heaps together a huge pile of assorted rubbish to be burned. The person preparing this huge bonfire probably gives little thought about the long-range transport (and consequences) of any contaminants that might be generated from that bonfire. This person simply has trash he or she no longer wants, and an easy solution is to burn it.

This pile of rubbish contains various elements: discarded rubber tires, old compressed gas bottles, assorted plastic containers, paper, oils and greases, wood, and old paint cans. The person burning it does not consider this hazardous material, though, just household trash. When the pile of rubbish burns, a huge plume of smoke forms and is carried away by a westerly wind. The fire-starter looks downwind and notices that the smoke disappears just a few miles over the property line. The dilution processes and the enormity of the atmosphere work together to dissipate and move away the smoke plume; the fire-starter does not give it a second thought. However, elevated levels of pollutants from many such fires may occur hundreds to thousands of miles downwind of the combination point sources producing such plumes. The result is that people living many miles from such pollution generators end up breathing contaminated air, transported over distance to their location.

Transport or dispersion estimates are determined by using distribution equations and air quality models. These dispersion estimates are typically valid for the layer of the atmosphere closest to the ground, where frequent changes occur in the temperature and distribution of the winds. These two variables have an enormous effect on how plumes are dispersed.

DISPERSION MODELS

Air quality dispersion models consist of a set of mathematical equations that interpret and predict pollutant concentration due to plume dispersal and impaction. They are essentially used to predict or describe the fate of airborne gases, particulate matter, and ground-level concentrations downwind of point sources. To determine the significance of air quality impact to a particular area, the first consideration is normal background concentrations, those pollutant concentrations from natural sources or distant, unidentified man-made sources. Each particular geographical area has a signature or background level of contamination considered to be an annual mean background concentration level of certain pollutants. An area, for example, might normally have a particulate matter reading of 30-40 μg/m³. If particulate matter readings are significantly higher than the background level, this suggests an additional source. To establish background contaminations for a particular source, air quality data related to that site and its vicinity must be collected and analyzed.

The USEPA recognized that in calculating the atmospheric dispersion of air pollutants, some means by which consistency could be maintained in air quality analysis had to be established. Thus, the USEPA promulgated two guidebooks to assist in modeling for air quality analyses: *Guidelines on Air Quality Models* (Revised) (1986) and *Industrial Source Complex (ISC) Dispersion Models User's Guide* (1986).

In performing dispersion calculations, particularly for health effect studies, the USEPA and other recognized experts in the field recommend following a four-step procedure:

1. Estimate the rate, duration, and location of the release into the environment.
2. Select the best available model to perform the calculations.
3. Perform the calculations and generate downstream concentrations, including lines of constant concentration (isopleths) resulting from the source emissions.
4. Determine what effect, if any, the resulting discharge has on the environment, including humans, animals, vegetation, and materials of construction.

These calculations often include estimates of the so-called vulnerability zones, that is, regions that may be adversely affected because of the emissions (Holmes et al., 1993).

Before beginning any dispersion determination activity, you must first determine the acceptable ground-level concentration of the waste pollutant(s). Local meteorological conduits and local topography must be considered, and having an accurate knowledge of the constituents of the waste gas and its chemical and physical properties is paramount.

Air quality models provide a relatively inexpensive means of determining compliance and predicting the degree of emission reduction necessary to attain ambient air quality standards. Under the 1977 Clean Air Act Amendments, the use of models is required for the evaluation of permit applications associated with permissible increments under the so-called Prevention of Significant Deterioration (PSD) requirements, which require localities "to protect and enhance"" air that is not contaminated (Godish, 1997).

Several dispersion models have been developed. Really equations, these models are mathematical descriptions of the meteorological transport and dispersion of air contaminants in a particular area, which permit estimates of contaminant concentrations, in plume from either a ground-level source or an elevated source (Carson and Moses, 1969). User-friendly modeling programs are available now that produce quick, accurate results from the operator's pertinent data.

There are four generic types of models: Gaussian, numerical, statistical, and physical. The *Gaussian* models use the Gaussian distribution equation and are widely used to estimate the impact of nonreactive pollutants. *Numerical* models are more appropriate than Gaussian models for area sources in urban locations that involve reactive pollutants, but numerical models require extremely detailed source and pollutant information and are not widely used. *Statistical* models are used when scientific information about the chemical and physical processes of a source is incomplete or vague and therefore make the use of either Gaussian or numerical models impractical. Lastly, *physical* models require fluid modeling studies or wind tunneling. This approach involves the construction of a scaled model and the observation of fluid flow around these models. This type of modeling is very complex and requires expert technical support. However, for large areas with complex terrain, stack downwash, complex flow conditions, or large buildings, this type of modeling may be the best choice.

As mentioned, selection of an air quality model for a particular air quality analysis is dependent on the type of pollutants being emitted, the complexity of the source, and the type of topography surrounding the facility. Some pollutants are formed by the combination of precursor pollutants. For example, ground-level ozone is formed when volatile organic compounds (VOCs) and nitrogen oxides (NO_x) react in the presence of sunlight. Models to predict ground-level ozone concentrations would use the emission rate of VOCs and NO_x as inputs. Also, some pollutants readily react once emitted into the atmosphere. These reactions deplete the concentrations of these pollutants and may need to be accounted for in the model. Source complexity also plays a role in model selection. Some pollutants may be emitted from short stacks that are subject to aerodynamic downwash. If this is the case, a model must be used that is capable of accounting for this phenomenon. Again, topography plays a major role in the dispersal of plumes and their air pollutants and must be considered in the selection of an air quality model. Elevated plumes may impact areas of high terrain. Elevated terrain heights may experience higher pollutant concentrations since they are closer to the plume centerline. A model that considers terrain heights should be used when elevated terrain exists.

This book's intent is not to develop each dispersion model in detail, but rather to recommend the one with the greatest applicability today. Probably the best atmospheric dispersion workbook for modeling published to date is that by D. B. Turner, *Workbook of Atmospheric Dispersion Estimates*, for the EPA, and most of the air dispersion models used today are based on the Pasquill-Gifford model.

MAJOR AIR POLLUTANTS

The most common and widespread anthropogenic pollutants currently emitted are sulfur dioxide (SO_2), nitrogen oxides (NO_x), carbon monoxide (CO), carbon dioxide (CO_2), volatile organic compounds (hydrocarbons), particulates, lead, and several toxic chemicals. Table 15.1 lists important air pollutants and their sources.

Recall that in the United States, the Environmental Protection Agency (EPA) regulates air quality under the Clean Air Act (CAA) and amendments that charged the federal government to develop uniform National Ambient Air Quality Standards

TABLE 15.1

Pollutant	Source
Sulfur and nitrogen oxides	From fossil fuel combustion
Carbon monoxide	Mostly from motor vehicles
Volatile organic compounds	From vehicles and industry
Ozone	From atmospheric reactions between nitrogen oxides and organic compounds

Source: USEPA, *Environmental Progress and Challenges*, 1988a, p. 13.

(NAAQS), which were discussed in Chapter 14. These were to include a dual requirement of primary standards (covering criteria pollutants) designed to protect health and secondary standards to protect public welfare. Primary standards were to be achieved by July 1975, and secondary standards in "a reasonable period of time." Pollutant levels protective of public welfare take priority over (and are more stringent than) those for public health; achievement of the primary health standard had immediate priority. In 1971 the USEPA promulgated NAAQS for six classes of air pollutants. Later, in 1978, an air quality standard was also promulgated for lead and the photochemical oxidant standard was revised to an ozone (O_3) standard (the ozone permissible level was increased). The particular matter standard was revised and redesignated the PM_{10} standard in 1987. This revision reflected the need for a PM standard based on particle sizes (≤ 10 μm) that have the potential for entering the respiratory tract and affecting human health. The National Ambient Air Quality Standards (with 1998 updates) are summarized in Table 14.1.

Thus, air pollutants were categorized into two groups: primary and secondary. Primary pollutants are emitted directly into the atmosphere where they exert an adverse influence on human health or the environment. Of particular concern are primary pollutants emitted in large quantities: carbon dioxide, carbon monoxide, sulfur dioxide, nitrogen dioxides, hydrocarbons, and particulate matter (PM). Once in the atmosphere, primary pollutants may react with other primary pollutants or atmospheric compounds such as water vapor to form secondary pollutants. A secondary pollutant that has received a lot of press and attention otherwise is acid precipitation, which is formed when sulfur or nitrogen oxides react with water vapor in the atmosphere.

SULFUR DIOXIDE (SO_2)

Sulfur enters the atmosphere in the form of corrosive *sulfur dioxide* (SO_2) gas. Sulfur dioxide is a colorless gas possessing the sharp, pungent odor of burning rubber. On a global basis, nature and anthropogenic activities produce sulfur dioxide in roughly equivalent amounts. Its natural sources include volcanoes, decaying organic matter, and sea spray, while anthropogenic sources include combustion of sulfur-containing coal and petroleum products and smelting of nonferrous ores. According to the World Resources Institute and International Institute for Environment and Development

(WRI and IIED, 1988–1989), in industrial areas much more sulfur dioxide comes from human activities than from natural sources. Sulfur-containing substances are often present in fossil fuels; SO_2 is a product of combustion that results from the burning sulfur-containing materials. The largest single source (65%) of sulfur dioxide is from the burning of fossil fuels to generate electricity. Thus, near major industrialized areas, it is often encountered as an air pollutant.

In the air, sulfur dioxide converts to sulfur trioxide (SO^3) and sulfate particles (SO_4). Sulfate particles restrict visibility and, in the presence of water, form sulfur acid (H_2SO_4), a highly corrosive substance that also lowers visibility. According to McKenzie and El-Ashry (1988), global output of sulfur dioxide has increased sixfold since 1900. Most industrial nations, however, have since (1975–1985) lowered sulfur dioxide levels by 20 to 60% by shifting away from heavy industry and imposing stricter emission standards. Major sulfur dioxide reductions have come from burning coal with lower sulfur content and from using less coal to generate electricity.

Two major environmental problems have developed in highly industrialized regions of the world, where the atmospheric sulfur dioxide concentration has been relatively high: sulfurous smog and acid rain. Sulfurous smog is the haze that develops in the atmosphere when molecules of sulfuric acid accumulate, growing in size as droplets until they become sufficiently large to serve as light scatterers. The second problem, acid rain, is precipitation contaminated with dissolved acids like sulfuric acid. Acid ran has posed a threat to the environment by causing certain lakes to become void of aquatic life.

NITROGEN OXIDES (NO_X)

There are seven oxides of nitrogen that are known to occur—NO, NO_2, NO_3, N_2O, N_2O_3, N_2O_4, and N_2O_5—but only two are important in the study of air pollution: nitric oxide (NO) and nitrogen dioxide (NO_2). Nitric oxide is produced by both natural and human actions. Soil bacteria are responsible for the production of most of the nitric oxide that is produced naturally and released to the atmosphere. Within the atmosphere, nitric oxide readily combines with oxygen to form nitrogen dioxide, and together, those two oxides of nitrogen are usually referred to as NO_x (nitrogen oxides). NO_x are formed naturally by lightening and by decomposing organic matter. Approximately 50% of anthropogenic NO_x are emitted by motor vehicles, and about 30% come from power plants, with the other 20% produced by industrial processes.

Scientists distinguish between two types of NO_x—thermal and fuel—depending on its mode of formation. Thermal NO_x are created when nitrogen and oxygen in the combustion air, such as those within internal combustion engines, are heated to a high enough temperature (above 1000 K) to cause nitrogen (N_2) and oxygen (O_2) in the air to combine. Fuel NO_x result from the oxidation (i.e., combines with oxygen in the air) of nitrogen contained within a fuel such as coal. Both types of NO_x generate nitric oxide first, and then when vented and cooled, a portion of nitric oxide is converted to nitrogen dioxide. Although thermal NO_x can be a significant contributor to total NO_x emissions, fuel NO_x are usually the dominant source, with approximately 50% coming from power plants (stationary sources), and the other half is released by automobiles (mobile sources).

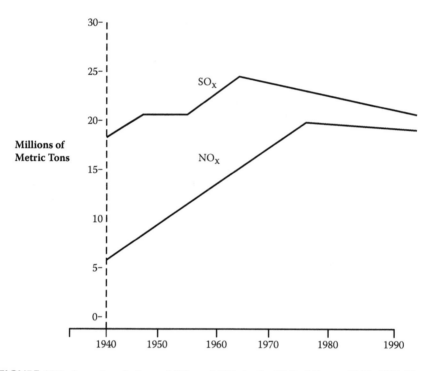

FIGURE 15.2 Annual emissions of SO_x and NO_x in the United States, 1940–1987. From USEPA, *National Air Pollutant Emissions Estimates*, 1989.

Nitrogen dioxide is more toxic than nitric oxide, and is a much more serious air pollutant. Nitrogen dioxide, at high concentrations, is believed to contribute to heart, lung, liver, and kidney damage. In addition, because nitrogen dioxide occurs as a brownish haze (giving smog its reddish brown color), it reduces visibility. When nitrogen dioxide combines with water vapor in the atmosphere, it forms nitric acid (HNO_3), a corrosive substance that, when precipitated out as acid rain, causes damage to plants and corrosion of metal surfaces.

NO_x rose in several countries and then leveled off or declined during the 1970s. During this same timeframe (see Figure 15.2), levels of nitrogen oxide have not dropped as dramatically as those of sulfur dioxide, primarily because a large part of total NO_x emissions comes from millions of motor vehicles, while most sulfur dioxide is released by a relatively small number of emission-controlled, coal-burning power plants.

CARBON MONOXIDE (CO)

Carbon monoxide is a colorless, odorless, tasteless gas formed when carbon in fuel is not burned completely; it is by far the most abundant of the primary pollutants, as Table 15.2 indicates. When inhaled, carbon monoxide gas restricts the blood's

TABLE 15.2
United States Emission Estimates, 1986 (10^{12} g/year)

Source	SO_x	NO^x	VOC	CO	Lead	PM
Transportation	0.9	8.5	6.5	42.6	0.0035	1.4
Stationary source fuel	17.2	10.0	2.3	7.2	0.0005	1.8
Industrial processes	3.1	0.6	7.9	4.5	0.0019	2.5
Solid waste disposal	0.0	0.1	0.6	1.7	0.0027	0.3
Miscellaneous	0.0	0.1	2.2	5.0	0.0000	0.8
Total	28.4	18.1	19.5	60.9	0.0086	6.8

Source: USEPA, *National Air Pollutant Emission Estimates 1940–1986*, Washington, DC: Environmental Protection Agency, 1988.

ability to absorb oxygen, causing angina, impaired vision, and poor coordination. Carbon monoxide has little direct effect on ecosystems, but has an indirect environmental impact via contributing to greenhouse effect and depletion of the earth's protective ozone layer.

The most important natural source of atmospheric carbon monoxide is the combination of oxygen with methane (CH_4), which is a product of the anaerobic decay of vegetation. (Anaerobic decay takes place in the absence of oxygen.) At the same time, however, carbon monoxide is removed from the atmosphere by the activities of certain soil microorganisms, so the net result is a harmless average concentration that is less than 0.12–15 ppm, in the northern hemisphere. Because stationary source combustion facilities are under much tighter environmental control than are mobile sources, the principal source of carbon monoxide that is caused by human activities is motor vehicle exhaust, which contributes to about 70% of all CO emissions in the United States (see Figure 15.3).

VOLATILE ORGANIC COMPOUNDS (HYDROCARBONS)

Volatile organic compounds (VOCs; also listed under the general heading of hydrocarbons) encompass a wide variety of chemicals that contain exclusively hydrogen and carbon. Emissions of volatile hydrocarbons from human resources are primarily the result of incomplete combustion of fossil fuels. Fires and the decomposition of matter are the natural sources. Of the VOCs that occur naturally in the atmosphere, methane (CH_4) is present at highest concentrations (approximately 1.5 ppm). But even at relatively high concentrations methane does not interact chemically with other substances and causes no ill health effects. However, in the lower atmosphere, sunlight causes VOCs to combine with other gases, such as NO_2, oxygen, and CO, to form secondary pollutants such as formaldehyde, ketones, ozone, peroxyacetyl nitrate (PAN), and other types of photochemical oxidants. These active chemicals can irritate the eyes and damage the respiratory system and damage vegetation.

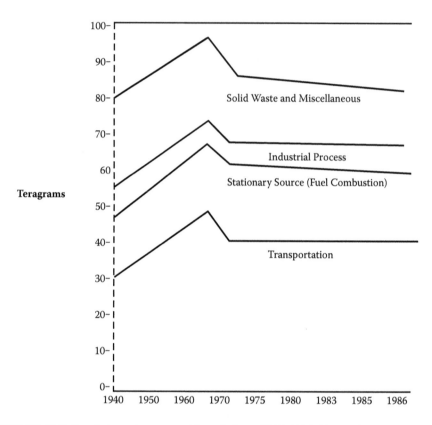

OZONE AND PHOTOCHEMICAL SMOG

By far the most damaging photochemical air pollutant is ozone (each ozone molecule contains three atoms of oxygen and thus is written O_3). Other photochemical oxidants (PAN, hydrogen peroxide (H_2O_2), and aldehydes) play minor roles. All of these are secondary pollutants because they are not emitted, but are formed in the atmosphere by photochemical reactions involving sunlight and emitted gases, especially NO_x and hydrocarbons.

Ozone is a bluish gas, about 1.6 times heavier than air, and relatively reactive as an oxidant. Ozone is present in a relatively large concentration in the stratosphere and is formed naturally by ultraviolet radiation. At ground level, ozone is a serious air pollutant; it has caused serious air pollution problems throughout the industrialized world, posing threats to human health and damaging foliage and building material.

According to MacKenzie and El-Ashry (1988), ozone concentrations in industrialized countries of North America and Europe are up to three times higher than the level at which damage to crops and vegetation begins. Ozone harms vegetation by damaging plant tissues, inhibiting photosynthesis, and increasing susceptibility to disease, drought, and other air pollutants.

In the upper atmosphere, where good (vital) ozone is produced, ozone is being depleted by anthropogenic emission of ozone-depleting chemicals on the ground. With this increase, concern has been raised over a potential upset of the dynamic equilibria among stratospheric ozone reactions, with a consequent reduction in ozone concentration. This is a serious situation because stratospheric ozone absorbs much of the incoming solar ultraviolet (UV) radiation. As a UV shield, ozone helps to protect organisms on the earth's surface from some of the harmful effects of this high-energy radiation. If not interrupted, UV radiation could cause serious damage, as disruption of genetic material, which could lead to increased rates of skin cancers and heritable problems.

In the mid-1980s a serious problem with ozone depletion became apparent. A springtime decrease in the concentration of stratospheric ozone (ozone holes) had been observed at high latitudes, most notably over Antarctica between September and November. Scientists strongly suspected that chlorine atoms or simple chlorine compounds may play a key role in this ozone depletion problem.

On rare occasions, it is possible for upper stratospheric ozone (good ozone) to enter the lower atmosphere (troposphere). Generally, this phenomenon only occurs during an event of great turbulence in the upper atmosphere. On rare incursions, atmospheric ozone reaches ground level for a short period of time. Most of the tropospheric ozone is formed and consumed by endogenous photochemical reactions, which are the result of the interaction of hydrocarbons, oxides of nitrogen, and sunlight, which produces a yellowish brown haze commonly called smog (Los Angeles–type smog).

Although the incursion of stratospheric ozone into the troposphere can cause smog formation, the actual formation of Los Angeles–type smog involves a complex group of photochemical interactions. These interactions are between anthropogenically emitted pollutants (NO and hydrocarbons) and secondarily produced chemicals (PAN, aldehydes, NO_2, and ozone). Note that the concentrations of these chemicals exhibit a pronounced diurnal pattern, depending on their rate of emission and the intensity of solar radiation and atmospheric stability at different times of the day (Freedman, 1989). This pattern is illustrated in Figure 15.4, for the important pollutant gases that contribute to Los Angeles–type smog.

If we look at Figure 15.4 and follow the timeline for the presence of various air pollutants in the atmosphere of Los Angeles, it is obvious that NO (emitted as NO_x) has a morning peak of concentration between 0600 and 0700, largely due to emissions from morning rush-hour vehicles. Hydrocarbons are emitted from both vehicles and refineries; they display a pattern similar to that of NO except that their peak concentration is slightly later. In bright sunlight the NO is photochemically oxidized to NO_2, resulting in a decrease in NO concentration and a peak of NO_2 at 0700–0900. Photochemical reactions involving NO_2 produce O atoms, which react with O_2 to form O_3. These result in a net decrease in NO_2 concentration and an increase in O_3 concentration, peaking between 1200 and 1500. Aldehydes, also formed photochemically, peak earlier than O_3. As the day proceeds, the various gases decrease in concentration as they are diluted by fresh air masses, or are consumed

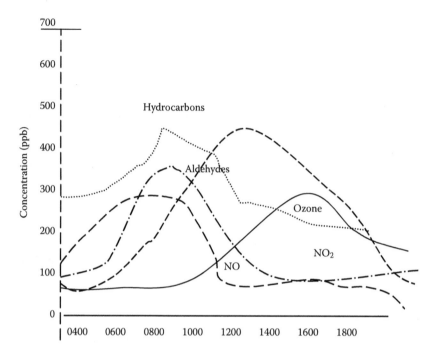

FIGURE 15.4 Average concentration of various air pollutants in the atmosphere of Los Angeles during days of eye irritation. Adaptation from Haagen-Smit Wayne, in Stern, A. C., ed., *Air Pollution*, 3rd ed., vol. 1, New York: Academic Press, 1976, pp. 235–88.

by photochemical reactions. This cycle is typical of an area that experiences photochemical smog and is repeated daily (Urone, 1976).

A tropospheric ozone budget for the northern hemisphere is shown in Table 15.3. The considerable range of the estimates reflects uncertainty in the calculation of the ozone fluxes. On average, stratospheric incursions account for about 18% of the total ozone influx to the troposphere, while endogenous photochemical production accounts for the remaining 82%. About 31% of the tropospheric ozone is consumed

TABLE 15.3
Tropospheric Ozone Budget (Northern Hemisphere)
(kg/ha-year)

Transport from stratosphere	13–20
Photochemical production	48–78
Destruction at ground	18–35
Photochemical destruction	48–55

Source: Adapted from Hov, O., *Ambio* 13, 73–79, 1984.

by oxidative reactions at vegetative and inorganic suffocates at ground level, while the other 69% is consumed by photochemical reactions in the atmosphere (Freedman, 1989).

CARBON DIOXIDE

Carbon-laden fuels, when burned, release carbon dioxide (CO_2) into the atmosphere. Much of this carbon dioxide is dissipated and then absorbed by ocean water, some is taken up by vegetation through photosynthesis, and some remains in the atmosphere. Today, the concentration of carbon dioxide in the atmosphere is approximately 350 ppm, and is rising at a rate of approximately 20 ppm every decade. The increasing rate of combustion of coal and oil has been primarily responsible for this occurrence, which may eventually have an impact on global climate.

PARTICULATE MATTER

Atmospheric particulate matter is defined as any dispersed matter, solid, or liquid in which the individual aggregates are larger than single small molecules, but smaller than about 500 µm. Particulate matter is extremely diverse and complex, since size and chemical composition, as well as atmospheric concentrations, are important characteristics (Masters, 1991).

A number of terms are used to categorize particulates, depending on their size and phase (liquid or solid). These terms are listed and described in Table 15.4.

TABLE 15.4
Atmospheric Particulates

Term	Description
Aerosol	General term for particles suspended in air
Mist	Aerosol consisting of liquid droplets
Dust	Aerosol consisting of solid particles that are blown into the air or produced from larger particles by grinding them down
Smoke	Aerosol consisting of solid particles or a mixture of solid and liquid particles produced by chemical reactions such as fires
Fume	Generally means the same as smoke, but often applies specifically to aerosols produced by condensation of hot vapors, especially of metals
Plume	The geometrical shape or form of the smoke coming out of a stack or chimney
Fog	Aerosol consisting of water droplets
Haze	Any aerosol, other than fog, that obscures the view through the atmosphere
Smog	Popular term originating in England to describe a mixture of smoke and fog; implies photochemical pollution

Dust, spray, forest fires, and the burning of certain types of fuels are among the sources of particulates in the atmosphere. Even with the implementation of stringent emission controls, which have worked to reduce particulates in the atmosphere, the U.S. Office of Technology Assessment (Postel, 1987) estimates that current levels of particulates and sulfates in ambient air may cause the premature death of 50,000 Americans every year.

LEAD

Lead is emitted to the atmosphere primarily from human sources, such as burning leaded gasoline, in the form of inorganic particulates. In high concentrations, lead can damage human health and the environment. Once lead enters an ecosystem, it remains there permanently. In humans and animals, lead can affect the neurological system and cause kidney disease. In plants, lead can inhibit respiration and photo-synthesis as well as block the decomposition of microorganisms. Since the 1970s, stricter emission standards have caused a dramatic reduction in lead output.

REFERENCES AND RECOMMENDED READING

Barker, J. R., and Tingey, D. T. 1991. *Air pollution effects on biodiversity.* New York: Van Nostrand Reinhold.
Carson, J. E., and Moses, H. 1969. The validity of several plume rise formulas. *J. Air Pol. Cont. Assoc.* 19:862.
Freedman, B., 1989. *Environmental ecology,* New York: Academic Press.
Godish, T. 1997. *Air quality.* 3rd ed. Boca Raton, FL: Lewis Publishers.
Holmes, G., Singh, B. R., and Theodore, L. 1993. *Handbook of environmental management and technology.* New York: John Wiley & Sons.
MacKenzie, J. J., and El-Ashry, T. 1988. *Ill winds: Airborne pollutant's toll on trees and crops.* Washington, DC: World Resource Institute.
Masters, G. M. 1991. *Introduction to environmental engineering and science.* Englewood Cliffs, NJ: Prentice Hall.
Peavy, H. S., Rowe, D. R., and Tchobanglous, G. 1985. *Environmental engineering.* New York: McGraw-Hill.
Postel, S. 1987. Stabilizing chemical cycles. In Brown, L. R., ed., *State of the world.* New York: Norton.
Spellman, F. R., and Whiting, N. 2006. *Environmental science and technology: Concepts and applications.* Rockville, MD: Government Institutes.
Urone, P. 1976. The primary air pollutants—Gaseous. Their occurrence, sources, and effects. In Stern, A. C., ed., *Air pollution.* Vol. 1. New York: Academic Press.
USEPA. 2005. *Basic air pollution meteorology.* www.epa.gov/apti (accessed January 18, 2008).

16 Air Pollution Control Technology

> There are two primary motivations behind the utilization of industrial air pollution control technologies. These are
>
> 1. They must be used because of legal or regulatory requirements
> 2. They are integral to the economical operation of an industrial process
>
> Although economists would point out that both of these motivations are really the same, that is, it is less expensive for an industrial user to operate with air pollution control than without, the distinction in application type is an important one.... In general, air pollution control is used to describe those applications that are driven by regulations and/or health considerations, while applications that deal with product recovery are considered process applications. Nevertheless, the technical issues, equipment design, operation, etc., will be similar if not identical. In fact, what differs between these uses is that the economics that affect the decision making process will often vary to some degree.
>
> **W. L. Heumann, 1997, p. xv**

INTRODUCTION

Chapters 13 to 15 set the foundation for the discussion presented in this chapter. Now that you have a clear picture of the problems that air pollution control is trying to solve, the time has come to examine the measures used to control it.

Two important factors related to the topic are presented in the opening sentence of this chapter's introductory quote: control technology and regulation. Neither is more important than the other. In fact, in many ways, they drive each other.

Air pollution control begins with regulation. Regulations (for example, to clean up, reduce, or eliminate a pollutant emission source), in turn, are generated because of certain community concerns. Buonicore et al. (1992) point out that regulations usually evolve around three considerations:

1. Legal limitations imposed for the protection of public health and welfare
2. Social limitations imposed by the community in which the pollution source is or is to be located
3. Economic limitations imposed by marketplace constraints

The engineer assigned to mitigate an air pollution problem must ensure that the design control methodology used will bring the source into full compliance with applicable regulations. To accomplish this feat, environmental engineers must first understand the problems and then rely heavily on technology to correct the situation. Various air pollution control technologies are available to environmental engineers or air pollution control practitioners working to mitigate air pollution source problems. By analyzing the problem carefully and applying the most effective method for the situation, the engineer or practitioner can ensure that a particular pollution source is brought under control and the responsible parties are in full compliance with regulations.

In this chapter, we discuss the various air pollution control technologies available to the environmental engineers and air pollution control practitioners in mitigating air pollution source problems.

AIR POLLUTION CONTROL: CHOICES

Assuming that the design engineer has a complete knowledge of the contaminant and the source, all available physical and chemical data on the effluent from the source, and the regulations of the control agencies involved, he or she must then decide which control methodology to employ. Since only a few control methods exist, the choice is limited. Control of atmospheric emissions from a process will generally consist of one of four methods depending on the process, types, fuels, availability of control equipment, etc. The four general control methods are (1) elimination of the process entirely or in part, (2) modification of the operation to a fuel that will give the desired level of emission, (3) installation of control equipment between the pollutant source and the receptor, and (4) relocation of the operation.

Tremendous costs are involved with eliminating or relocating a complete process, which makes either of these choices the choice of last resort. Let us take a look at the first and last control methods first. Eliminating a process is no easy undertaking, especially when the process to be eliminated is that for which the facility exists. Relocation is not always an answer either. Consider the real-life situation presented below.

Cedar Creek Composting (CCC) facility was built in 1970. A 44-acre site designed to receive and process compost wastewater biosolids from six local wastewater treatment plants, CCC composted biosolids at the rate of 17.5 dry tons per day. CCC used the aerated static pile (ASP) method to produce pathogen-free, humus-like material that can be beneficially used as an organic soil amendment. The final compost product was successfully marketed under a registered trademark name.

Today, Cedar Creek Composting facility is no longer in operation. The site was shut down in early 1997. From an economic point of view, CCC was highly successful. When a fresh pile of compost had completed the entire composting process (including curing), dump truck after dump truck would line the street outside the main gate, waiting in the hope to buy a load of the popular product. Economics was not the problem. In fact, CCC could not produce enough compost fast enough to satisfy the demand.

What was the problem? The answer to this is actually twofold: social and then eventually legal. The first problem was social limitations imposed by the community in which the compost site was located. In 1970, the 44 acres CCC occupied were located in an out-of-town, rural area. CCC's only neighbor was a regional, small airport on its eastern border. CCC was completely surrounded by woods on the other three sides. The nearest town was 2 miles away. But by the mid-1970s, things started to change.

Population growth and its accompanying urban sprawl quickly turned forested lands into housing complexes and shopping centers. CCC's western border soon became the site of a two-lane road that was upgraded to four and then six lanes. CCC's northern fence separated it from a mega-shopping mall. On the southern end of the facility, acres of houses, playgrounds, swimming pools, tennis courts, and a golf course were built. CCC became an island surrounded by urban growth. Further complicating the situation was the airport; it expanded to the point that by 1985, three major airlines used the facility.

CCC's ASP composting process was not a problem before the neighbors moved in. We all know dust and odor control problems are not problems until the neighbors complain—and complain they did. CCC was attacked from all four sides. The first complaints came from the airport. The airport complained that dust from the static piles of compost was interfering with air traffic control.

The new, expanded highway brought several thousand new commuters right up alongside CCC's western fence line. Commuters started complaining anytime the compost process was in operation; they complained primarily about the odor—that thick, earthy smell permeated everything.

After the enormous housing project was completed and people took up residence there, complaints were raised on a daily basis. The new home owners complained about the earthy odor and the dust that blew from the compost piles onto their properties anytime they were downwind from the site. The shoppers at the mall also complained about the odor.

City hall received several thousand complaints over the first few months before it took any action. The city environmental engineer was told to approach CCC's management and see if some resolution to the problem could be effected. CCC management listened to the engineer's concerns but stated that there was not a whole lot that the site could do to rectify the problem.

As you might imagine, this was not the answer the city fathers were hoping to get. Feeling the increasing pressure from local inhabitants, commuters, shoppers, and airport management people, the city brought the local state representatives into the situation. The two state representatives for the area immediately began a campaign to close down the CCC facility.

CCC was not powerless in this struggle—after all, CCC was there first, right? The developers and the people in those new houses did not have to buy land right next to the facility—right? Besides, CCC had the USEPA on its side. CCC was taking a waste product no one wanted, one that traditionally ended up in the local landfill (taking up valuable space), and turning it into a beneficial reuse product. CCC was helping to conserve and protect the local environment, a noble endeavor.

The city politicians did not really care about noble endeavors, but they did care about the concerns of their constituents, the voters. They continued their assault through the press, electronic media, legislatively, and by any other means they could bring to bear.

CCC management understood the problem and felt the pressure. They had to do something, and they did. Their environmental engineering division was assigned the task of coming up with a plan to mitigate not only CCC's odor problem, but also its dust problem. After several months of research and a pilot study, CCC's environmental engineering staff came up with a solution. The solution included enclosing the entire facility within a self-contained structure. The structure would be equipped with a state-of-the-art ventilation system and two-stage odor scrubbers. The engineers estimated that the odor problem could be reduced by 90% and the dust problem reduced by 98.99%. CCC management thought they had a viable solution to the problem and were willing to spend the $5.2 million to retrofit the plant.

After CCC presented their mitigation plan to the city council, the council members made no comment but said that they needed time to study the plan. Three weeks later, CCC received a letter from the mayor stating that CCC's efforts to come up with a plan to mitigate the odor and dust problems at CCC were commendable and to be applauded but were unacceptable.

From the mayor's letter, CCC could see that the focus of attack had now changed from a social to a legal issue. The mayor pointed out that he and the city fathers had a legal responsibility to ensure the good health and well-being of local inhabitants and that certain legal limitations would be imposed and placed on the CCC facility to protect their health and welfare.

Compounding the problem was the airport. Airport officials also rejected CCC's plan to retrofit the compost facility. Their complaint (written on FAA official paper) stated that the dust generated at the compost facility was hazarding flight operations, and even though the problem would be reduced substantially by engineering controls, the chance of control failure was always possible, and then an aircraft could be endangered. From the airport's point of view, this was unacceptable.

Several years went by, with local officials and CCC management contesting each other on the plight of the compost facility. In the end, CCC management decided they had to shut down their operation and move to another location, so they closed the facility.

After shutdown, CCC management staff immediately started looking for another site to build a new wastewater biosolids-to-compost facility. They are still looking. To date, their search has located several pieces of property relatively close to the city (but far enough away to preclude any dust and odor problems), but they have had problems finalizing any deal. Buying the land is not the problem—getting the required permits from various county agencies to operate the facility is. CCC officials were turned down in each and every case. The standard excuse? Not in my backyard. Have you heard this phrase before? It is so common now, it is usually abbreviated—NIMBY. Whether back in the day or at present, NIMBY is alive and well.

To this very year (2008), CCC officials are still looking for a location for their compost facility; they are not all that optimistic about their chances of success in this matter.

The second pollution control method—modification of the operation to a fuel that will give the desired level of emission—often looks favorable to those who have weighed the high costs associated with air pollution control systems. Modifying the process to eliminate as much of the pollution problem as possible at the source is generally the first approach to be examined.

Again, often the easiest way to modify a process for air pollution control is to change the fuel. If a power plant, for example, emits large quantities of sulfur dioxide and fly ash, conversion to cleaner-burning natural gas is cheaper than installing the necessary control equipment to reduce the pollutant emissions to permitted values.

Changing from one fuel to another, however, causes its own problems related to costs, availability, and competition. Today's fuel prices are high, and no one counts on the trend reversing. Finding a low-sulfur fuel is not easy, especially since many industries own their own dedicated supplies (which are not available for use in other industries). With regulation compliance threatening everyone, everyone wants their share of any available low-cost, low-sulfur fuel. With limited supplies available, the law of supply and demand takes over and prices go up.

Some industries employ other process modification techniques. These may include evaluation of alternative manufacturing and production techniques, substitution of raw materials, and improved process control methods (Buonicore et al., 1992).

When elimination of the process entirely or in part, relocation of the operation, or modification of the operation to a fuel that will give the desired level of emission is not possible, the only alternative control method left is installation of control equipment between the pollutant source and the receptor (the purpose of installing pollution control equipment or a control system, obviously, is to remove the pollution from the polluted carrier gas). To accomplish this, the polluted carrier gas must pass through a control device or system, which collects or destroys the pollutant and releases the cleaned carrier gas to the atmosphere (Boubel et al., 1994). The rest of this chapter will focus on these air pollution control equipment devices and systems.

AIR POLLUTION CONTROL EQUIPMENT AND SYSTEMS

Several considerations must be factored into any selection decision for air pollution control equipment or systems. Careful consideration must be given to costs. No one ever said air pollution equipment/systems were inexpensive—they are not. Obviously, the equipment/system must be designed to comply with applicable regulatory emission limitations. The operational and maintenance history/record (costs of energy, labor, and repair parts should also be factored in) of each equipment/system must be evaluated. Remember, emission control equipment must be operated on a continual basis, without interruptions. Any interruption could be subject to severe regulatory penalties, which could again be quite costly.

Probably the major factor to consider in the equipment/system selection process is what type of pollutant or pollutant stream is under consideration. If the pollutant is conveyed in a carrier gas, for example, factors such as carrier gas pressure, temperature, viscosity, toxicity, density, humidity, corrosiveness, and inflammability must all be considered before any selection is made. Other important factors must also be considered when selecting air pollution control equipment. Many of the general factors are listed in Table 16.1.

In addition to those factors listed in Table 16.1, process considerations dealing with gas flow rate and velocity, pollutant concentration, allowable pressure drop, and the variability of gas and pollutant flow rates, including temperature, must all be considered.

The type of pollutant is also an important factor that must be taken into consideration: gaseous or particulate. Certain pertinent questions must be asked and answered. If the pollutant, for example, is gaseous, how corrosive, inflammable, reactive, and toxic is it? After these factors have been evaluated, the focus shifts to the selection of the best air pollution control equipment/system—affordable, practical, and permitted by regulatory requirements—depending, of course, on the type of pollutant to be removed.

In the following sections two types of pollutants (dry particulates and gaseous pollutants) and the various air pollution control equipment/processes available for their removal are discussed.

TABLE 16.1
Factors in Selecting Air Pollution Control
Equipment/Systems

1.	Best available technology (BAT)
2.	Reliability
3.	Lifetime and salvage value
4.	Power requirements
5.	Collection efficiency
6.	Capital cost, including operation and maintenance costs
7.	Track record of equipment/system and manufacturer
8.	Space requirements and weight
9.	Power requirements
10.	Availability of space, parts and manufacturer's representatives

REMOVAL OF DRY PARTICULATE MATTER

Constituting a major class of air pollutants, particulates have a variety of shapes and sizes, and as either liquid droplet or dry dust, they have a wide range of physical and chemical characteristics. Dry particulates are emitted from a variety of different sources, including both combustion and noncombustion sources in industry, mining, construction activities, incinerators, and internal combustion engines. Dry particulates are also emitted from natural sources—volcanoes, forest fires, pollen, and windstorms.

All particles and particulate matter exhibit certain important characteristics, which, along with process conditions, must be considered in any engineering strategy to separate and remove them from a stream of carrier gas. Particulate size range and distribution, particle shape, corrosiveness, agglomeration tendencies, abrasiveness, toxicity, reactivity, inflammability, and hygroscopic tendencies must all be examined in light of equipment limitations.

In an air pollution control system, particulates are separated from the gas stream by application of one or more forces, in gravity settlers, centrifugal settlers, fabric filters, electrostatic precipitators, or wet scrubbers. The particles are then collected and removed from the system.

As mentioned earlier in the text, when a flowing fluid (engineering and science applications consider both liquid and gaseous states as a fluid) approaches a stationary object such as a metal plate, a fabric thread, or a large water droplet, the fluid flow will diverge around that object. Particles in the fluid (because of inertia) will not follow stream flow exactly, but will tend to continue in their original directions. If the particles have enough inertia and are located close enough to the stationary object, they will collide with the object and can be collected by it. This is an important phenomenon.

Particles are collected by impaction, interception, and diffusion. *Impaction* occurs when the center of mass of a particle that is diverging from the fluid strikes a stationary object. *Interception* occurs when the particle's center of mass closely misses the object, but because of its finite size, the particle strikes the object. *Diffusion* occurs when small particulates happen to diffuse toward the object while passing near it. Particles that strike the object by any of these means are collected—if short-range forces (chemical, electrostatic, and so forth) are strong enough to hold them to the surface (Cooper and Alley, 1990).

Control technologies for particles focus on capturing the particles emitted by a pollution source. Several factors must be considered before choosing a particulate control device. Typically, particles are collected and channeled through a duct or stack. The characteristics of the particulate exhaust stream affect the choice of the control device. These characteristics include the range of particle sizes, the exhaust flow rate, the temperature, the moisture content, and various chemical properties, such as explosiveness, acidity, alkalinity, and flammability.

The most commonly used control devices for controlling particulate emissions include: gravity settlers, cyclones, electrostatic precipitators, wet (venturi) scrubbers, and baghouse (fabric filters). In many cases, more than one of these devices is used in a series to obtain the desired removal efficiencies. For example, a settling chamber can be used to remove larger particles before a pollutant stream enters an electrostatic precipitator. In this section we will briefly introduce each of the major types of particulate control equipment and point out their advantages and disadvantages.

GRAVITY SETTLERS

Gravity settlers (or settling chambers) have long been used by industry for removing solid and liquid waste materials from gaseous streams. Simply constructed (see Figure 16.1), a gravity settler is actually nothing more than an enlarged chamber in which the horizontal gas velocity is reduced, allowing large particles to settle out of the gas by gravity and be recollected in hoppers. Gravity settlers have the advantage of having low initial cost and are relatively inexpensive to operate—there is not a lot to go wrong. However, because settling chambers are effective in removing only larger particles, they have relatively low efficiency, especially for removal of small

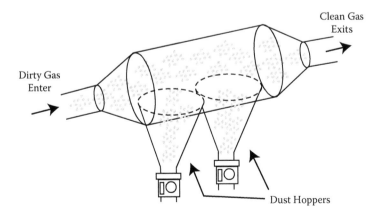

FIGURE 16.1 Gravity settler. From USEPA (2006).

particles (<50 μm). Thus, gravity settlers are used in conjunction with a more efficient control device. In addition, although simple in design, gravity settlers require a large space for installation.

CYCLONE COLLECTORS

The cyclone (or centrifugal) collector provides a low-cost, low-maintenance method of removing larger particulates from a gas stream. The cyclone removes particles by inertia separation, causing the entire gas stream to flow in a spiral pattern inside a tube, and is the collector of choice for removing particles greater than 10 μm in diameter. By centrifugal force, the larger particles move outward and collide with the narrowing wall of the tube. The particles slide down the wall and fall to the bottom of the cone, where they are removed. The cleaned gas flows out the top of the cyclone (see Figure 16.2). Along with their relatively low construction costs, cyclones need relatively small space requirements for installation. However, cyclones are efficient in removing large particles, but are not as efficient with smaller particles, especially particles below 10 μm in size, and they do not handle sticky materials well. For this reason, they are used with other particulate control devices. The most serious

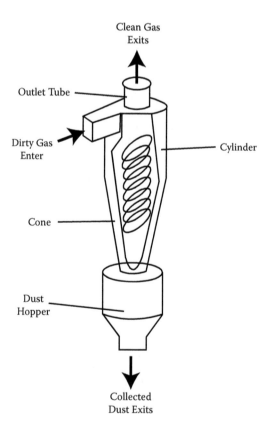

FIGURE 16.2 Cyclone. From USEPA (2006).

problems encountered with cyclones are with airflow equalization and their tendency to plug. Cyclones have been used successfully at feed and grain mills, cement plants, fertilizer plants, petroleum refineries, and other applications involving large quantities of gas containing relatively large particles.

Did You Know?

Because the particulate control devices discussed above do not destroy solid particles, proper disposal of the collected material is needed. Collected solid particles are most often disposed in a landfill. Wastewater generated by scrubber must be sent to a wastewater treatment facility. When possible, collected particle matter is recycled and reused.

ELECTROSTATIC PRECIPITATORS (ESPs)

An electrostatic precipitator (ESP) is a particle control device that uses electrical forces to move the particles out of the flowing gas stream and onto collector plates. ESPs are usually used to remove small particles from moving gas streams at high collection efficiencies. Widely used in power plants for removing fly ash from the gases prior to discharge, an electrostatic precipitator applies electrical force to separate particles from the gas stream. A high voltage drop is established between electrodes, and particles passing through the resulting electrical field acquire a charge. The charged particles are attracted to and collected on an oppositely charged plate, and the cleaned gas glows through the device. Periodically, the plates are cleaned by rapping to shake off the layer of dust that accumulates, and the dust is collected in hoppers at the bottom of the device (see Figure 16.3). Although electrostatic precipitators have the advantages of low operating costs, capability for operation at high-temperature

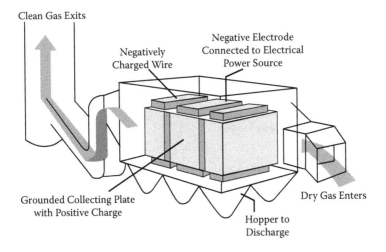

FIGURE 16.3 Electrostatic precipitator. From USEPA (2006).

applications (to 1,300°F), low pressure drop, and the attaining of extremely high particulate (coarse and fine) collection efficiencies, they have the disadvantages of high capital costs and space requirements. The removal efficiencies for ESPs are highly variable; however, for very small particles alone the removal efficiency is about 99%. Typical ESP applications include use in industrial and utility boilers, cement plants, steel mills, petroleum refineries, municipal waste incinerators, hazardous waste incinerators, Kraft pulp and paper mills, and lead, zinc, and copper smelters.

Wet (Venturi) Scrubbers

Wet scrubbers (or collectors) have found widespread use in cleaning contaminated gas streams (e.g., foundry dust emissions, acid mists, and furnace fumes) because of their ability to effectively remove particulate and gaseous pollutants. Wet scrubbers vary in complexity from simple spray chambers to remove coarse particles to high-efficiency systems (venturi types) to remove fine particles. Whichever system is used, operation employs the same basic principles of inertial impingement or impaction and interception of dust particles by droplets of water. The larger, heavier water droplets are easily separated from the gas by gravity. The solid particles can then be independently separated from the water, or the water can be otherwise treated before reuse or discharge. Increasing either the gas velocity or the liquid droplet velocity in a scrubber increases the efficiency because of the greater number of collisions per unit time. For the ultimate in wet scrubbing, where high collection efficiency is desired, the venturi scrubber is used. The venturi operates at extremely high gas and liquid velocities with a very high pressure drop across the venturi throat. The reduced velocity at the expanded section of the throat allows the droplets of water containing the particles to drop out of the gas stream. Venturi scrubbers such as the one shown in Figure 16.4 are most efficient for removing particulate matter in the size range of 0.5 to 5 μm, with removal efficiencies of up to 99%, which makes them especially effective for the removal of submicron particulates associated with smoke and fumes.

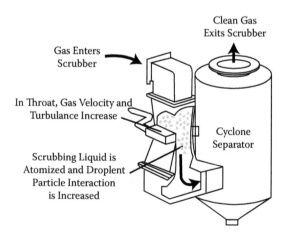

FIGURE 16.4 Venturi scrubber. From USEPA (2006).

TABLE 16.2
Advantages and Disadvantages of Wet Scrubbers

Advantages	Disadvantages
Small space requirements	Corrosion problems
No secondary dust sources	High power requirements
Handles high-temperature, high-humidity gas streams	Water disposal problems
Minimal fire and explosion hazards	Difficult product recovery
Ability to collect both gases and particles	Meteorological problems (plume = fog)

Source: USEPA (2006).

Although wet scrubbers require relatively small space requirements, can remove both gases and particles, can neutralize corrosive gases, have low capital cost, and can handle high-temperature, high-humidity gas streams, their power and maintenance costs are relatively high, they may create wastewater disposal problems, their corrosion problems are more severe than those of dry systems, and the final product they produce is collected wet. Table 16.2 summarizes these advantages and disadvantages. Wet scrubbers have been used in a variety of industries, such as acid plants, fertilizer plants, steel mills, asphalt plants, and larger power plants.

BAGHOUSE (FABRIC) FILTERS

Baghouse filters (or fabric filters) are the most commonly used air pollution control filtration system. In much the same manner as the common vacuum cleaner, fabric filter material, capable of removing most particles as small as 0.5 μm and substantial quantities of particles as small as 0.1 μm, is formed into cylindrical or envelope bags and suspended in the baghouse (see Figure 16.5). The particulate-laden gas stream is forced through the porous fabric filter, and as the air passes through the fabric, particulates accumulate on the cloth, providing a cleaned airstream. As particulates build up

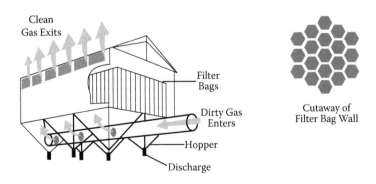

FIGURE 16.5 Fabric filter. From USEPA (2006).

on the inside surfaces of the bags, the pressure drop increases. Before the pressure drop becomes too severe, the bags must be relieved of some of the particulate layer. The particulates are periodically removed from the cloth by shaking or reversing the airflow.

The selection of the fiber material and fabric construction is important to baghouse performance. The fiber material from which the fabric is made must have adequate strength characteristics at the maximum gas temperature expected and adequate chemical compatibility with both the gas and the collected dust. One disadvantage of the fabric filter is that high-temperature gases often have to be cooled before contacting the filter medium.

Fabric filters are used in the power generation, incineration, chemical, steel, cement, food, pharmaceutical, metal working, aggregate, and carbon black industries.

Fabric filters are relatively simple to operate, provide high overall collection efficiencies up to 99%+, and are very effective in controlling submicrometer particles, but they do have limitations. These include relatively high capital costs, high maintenance requirements (bag replacement, etc.), high space requirements, and flammability hazards for some dusts.

REMOVAL OF GASEOUS POLLUTANTS: STATIONARY SOURCES

In the removal of gaseous air pollutants, the principal gases of concern are the sulfur oxides (SO_x), carbon oxide (CO_x), nitrogen oxides (NO_x), organic and inorganic acid gases, and hydrocarbons (HC). The most common method for controlling gaseous pollutants is the addition of control devices to recover or destroy a pollutant. Four major treatment processes (add-ons) currently available for control of these and other gaseous emissions are absorption, adsorption, condensation, and combustion (incineration).

The decision of which single or combined air pollution control technique to use for stationary sources is not always easy. Gaseous pollutants can be controlled by a wide variety of devices, and choosing the most cost-effective, most efficient units requires careful attention to the particular operation for which the control devices are intended. Specifically, the choice of control technology depends on the pollutants to be removed, the removal efficiency required, pollutant and gas stream characteristics, and specific characteristics of the site (EPA, 2006). Absorption, adsorption, and condensation all are recovery techniques, while incineration involves the destruction of the pollutant.

In making the difficult and often complex decision of which air pollution control technology to employ, it is helpful to follow guidelines based on experience and set forth by Buonicore and Davis (1992) in the prestigious engineering text *Air Pollution Engineering Manual*. Table 16.3 summarizes these.

ABSORPTION

Absorption (or scrubbing) is a major chemical engineering unit operation that involves bringing contaminated effluent gas into contact with a liquid absorbent so that one or more constituents of the effluent gas are selectively dissolved into a relatively nonvolatile liquid.

Absorption units are designed to transfer the pollutant from a gas phase to a liquid phase (water is the most commonly used absorbent liquid). The absorption

TABLE 16.3
Comparison of Air Pollution Control Technologies

Treatment Technology	Concentration and Efficiency	Comments
Incineration	(<100 ppmv) 90–95% efficiency (>100 ppmv) 95–99% efficiency	Incomplete combustion may require additional controls
Carbon adsorption	(>200 ppmv) 90%+ efficiency (>1,000 ppmv) 95%+ efficiency	Recovered organics may need additional treatment—can increase cost
Absorption	(<200 ppmv) 90–95% efficiency (>200 ppmv) 95%+ efficiency	Can blowdown stream be accommodated at site?
Condensation	(>2,000 ppmv) 80%+ efficiency	Must have low temperature or high pressure for efficiency

Note: Typically, only incineration and absorption technologies can achieve greater than 99% gaseous pollutant removal consistently.

unit accomplishes this by providing intimate contact between the gas and the liquid, providing optimum diffusion of the gas into the solution. The actual removal of a pollutant from the gas stream takes place in three steps: (1) diffusion of the pollutant gas to the surface of the liquid, (2) transfer across the gas/liquid interface, and (3) diffusion of the dissolved gas away from the interface into the liquid (Davis and Cornwell, 1991). Absorption is commonly used to recover products or to purify gas streams that have high concentrations of organic compounds. Absorption equipment is designed to get as much mixing between the gas and liquid as possible.

Several types of absorbers are available, including spray chambers (and towers or columns), plate or tray towers, packed towers, and venturi scrubbers. Pollutant gases commonly controlled by absorption include sulfur dioxide, hydrogen sulfide, hydrogen chloride, chlorine, ammonia, and oxides of nitrogen.

Absorbers are often referred to as scrubbers, and there are various types of absorption equipment. The principal types of gas absorption equipment include spray towers, packed columns, spray chambers, and venturi scrubbers. The two most common absorbent units in use today are the plate and packed tower systems. Plate towers contain perforated horizontal plates or trays designed to provide large liquid-gas interfacial areas. The polluted airstream is usually introduced at one side of the bottom of the tower or column and rises up through the perforations in each plate; the rising gas prevents the liquid from draining through the openings rather than through a downpipe. During continuous operation, contact is maintained between air and liquid, allowing gaseous contaminants to be removed, with clean air emerging from the top of the tower.

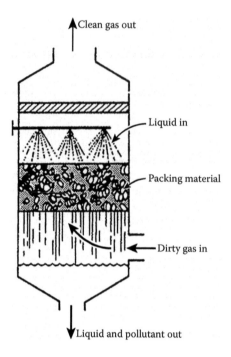

FIGURE 16.6 Typical countercurrent-flow packed tower. From USEPA (1971).

The packed tower scrubbing system (see Figure 16.6) is by far the most commonly used for the control of gaseous pollutants in industrial applications, where it typically demonstrates a removal efficiency of 90 to 95%. Usually configured in vertical fashion (Figure 16.6), the packed tower is literally packed with devices (see Figure 16.7) of

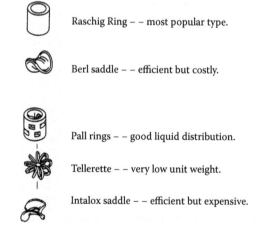

FIGURE 16.7 Various packing used in packed tower scrubbers. Adaptation from American Industrial Hygiene Association (1968).

large surface-to-volume ratio and a large void ratio that offers minimum resistance to gas flow. In addition, packing should provide even distribution of both fluid phases, be sturdy enough to support them in the tower, and be low cost, available, and easily handled (Hesketh, 1991).

The flow through a packed tower is typically countercurrent, with gas entering at the bottom of the tower and liquid entering at the top. Liquid flows over the surface of the packing in a thin film, affording continuous contact with the gases.

Though highly efficient for removal of gaseous contaminants, packed towers may create wastewater disposal problems (converting an air pollution problem to a water pollution problem), become easily clogged when gases with high particulate loads are introduced, and have relatively high maintenance costs.

Did You Know?

When a gas or vapor is brought into contact with a solid, part of it is taken up by the solid. The molecules that disappear from the gas either enter the inside of the solid or remain on the outside attached to the surface. The former phenomenon is termed absorption (or dissolution) and the latter adsorption.

ADSORPTION

Adsorption is a mass transfer process that involves passing a stream of effluent gas through the surface of prepared porous solids (adsorbents). The surfaces of the porous solid substance attract and hold (bind) the gas (the adsorbate) by either physical or chemical adsorption. In physical adsorption (a readily reversible process), a gas molecule adheres to the surface of the solid because of an imbalance of electron distribution. In chemical adsorption (not readily reversible), once the gas molecule adheres to the surface, it reacts chemically with it.

Several materials possess adsorptive properties. These include activated carbon, alumina, bone char, magnesia, silica gel, molecular sieves, strontium sulfate, and others. The most important adsorbent for air pollution control is activated charcoal. Activated carbon is the universal standard for purification and removal of trace organic contaminants from liquid and vapor streams. The surface area of activated charcoal will preferentially adsorb hydrocarbon vapors and odorous organic compounds from an airstream.

In an adsorption system (in contrast to the absorption system where the collected contaminant is continuously removed by flowing liquid), the collected contaminant remains in the adsorption bed. The most common adsorption system is the fixed-bed adsorber, which can be contained in either a vertical or horizontal cylindrical shell. The adsorbent (usually activated carbon) is arranged in beds or trays in layers about 0.5 inch thick. Multiple beds may be arranged as shown in Figure 16.8. In multiple-bed systems, one or more beds are adsorbing vapors, while the other bed is being regenerated.

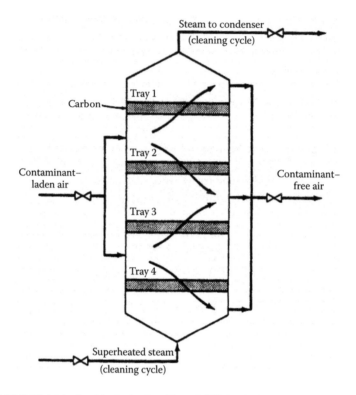

FIGURE 16.8 Multiple fixed-bed adsorber. From USEPA (1973).

The efficiency of most adsorbers is near 100% at the beginning of the operation and remains high until a breakpoint or breakthrough occurs. When the adsorbent becomes saturated with adsorbate, contaminant begins to leak out of the bed, signaling that the adsorber should be renewed or regenerated. By regenerating the carbon bed, the same activated carbon particles can be used again and again.

Although adsorption systems are high-efficiency devices that may allow recovery of product, have excellent control and response to process changes, and have the capability of being operated unattended, they also have some disadvantages, including the need for exotic, expensive extraction schemes if product recovery is required, relatively high capital cost, and gas stream prefiltering needs (to remove any particulate capable of plugging the adsorbent bed).

CONDENSATION

Condensation is a process by which volatile gases are removed from the contaminant stream and changed into a liquid. In air pollution control, a condenser can be used in two ways: either for pretreatment to reduce the load problems with other air pollution control equipment or for effectively controlling contaminants in the form of gases and vapors.

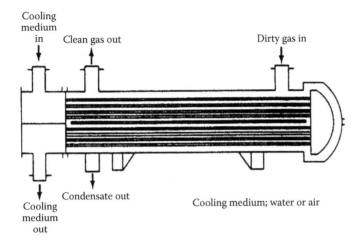

FIGURE 16.9 Surface condenser. From USEPA (1971).

Condensers condense vapors to liquid phase by either increasing the system pressure without a change in temperature, or decreasing the system temperature to its saturation temperature without a pressure change. Condensation is affected by the composition of the contaminant gas stream. When gases are present in the stream that condense under different conditions, condensation is hindered.

Condensers are widely used to recover valuable products in a waste stream. Condensers are simple, relatively inexpensive devices that normally use water or air to cool and condense a vapor stream. Condensers are typically used as pretreatment devices. They can be used ahead of adsorbers, absorbers, and incinerators to reduce the total gas volume to be treated by more expensive control equipment.

There are two basic types of condensation equipment used for pollution control—surface and contact condensers. A surface condenser is normally a shell-and-tube heat exchanger (see Figure 16.9). A surface condenser uses a cooling medium of air or water where the vapor to be condensed is separated from the cooling medium by a metal wall. Coolant flows through the tubes, while the vapor is passed over and condenses on the outside of the tubes, and drains off to storage (EPA, 2006).

In a contact condenser (which resembles a simple spray scrubber), the vapor is cooled by spraying liquid directly on the vapor stream (see Figure 16.10). The cooled vapor condenses, and the water and condensate mixture are removed, treated, and disposed of.

In general, contact condensers are less expensive, more flexible, and simpler than surface condensers, but surface condensers require much less water and produce many times less wastewater that must be treated than do contact condensers. Removal efficiencies of condensers typically range from 50% to more than 95%, depending on design and applications. Condensers are used in a wide range of industrial applications, including petroleum refining, petrochemical manufacturing, basic chemical manufacturing, dry cleaning, and degreasing.

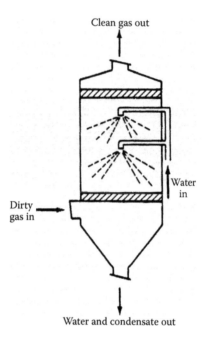

Clean gas out

Water
in

Dirty
gas in

Water and condensate out

FIGURE 16.10 Contact condenser. From USEPA (1971).

COMBUSTION (INCINERATION)

Even though combustion (or incineration) is a major source of air pollution, it is also, if properly operated, a beneficial air pollution control system in which the objective is to convert certain air contaminants (usually CO and hydrocarbons) to innocuous substances such as carbon dioxide and water (EPA, 2006).

Combustion is a chemical process defined as rapid, high-temperature gas-phase oxidation. The combustion equipment used to control air pollution emissions is designed to push these oxidation reactions as close to complete combustion as possible, leaving a minimum of unburned residue. The operation of any combustion operation is governed by four variables: oxygen, temperature, turbulence, and time. For complete combustion to occur, oxygen must be available and put into contact with sufficient temperature (turbulence), and held at this temperature for a sufficient time. These four variables are not independent—changing one affects the entire process.

Depending upon the contaminant being oxidized, equipment used to control waste gases by combustion can be divided into three categories: direct-flame combustion (or flaring), thermal combustion (afterburners), or catalytic combustion. Choosing the proper device depends on many factors, including type of hazardous contaminants in the waste stream, concentration of combustibles in the stream, process flow rate, control requirements, and an economic evaluation.

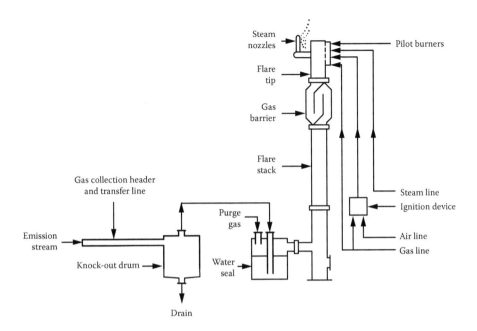

FIGURE 16.11 Schematic of steam-assisted flare system. From USEPA (1986).

DIRECT-FLAME COMBUSTION (FLARING)

Direct-flame combustion devices (flares) are the most commonly used air pollution control devices by which waste gases are burned directly (with or without the addition of a supplementary fuel). Common flares include steam-assisted, air-assisted, and pressure head types. Studies conducted by the EPA have shown that the destruction efficiency of a flare is about 98%. Flares are normally elevated from 100 to 400 feet to protect the surroundings from heat and flames. Often designed for steam injection at the flare top, flares commonly use steam in this application because it provides sufficient turbulence to ensure complete combustion, which prevents production of visible smoke or soot. Flares are also noisy, which can cause problems for adjacent neighborhoods, and some flares produce oxides of nitrogen, thus creating a new air pollutant. Figure 16.11 shows a steam-assisted flare system commonly used in industry.

THERMAL COMBUSTION (AFTERBURNERS)

The thermal incinerator or afterburner is usually the unit of choice in cases where the concentration of combustible gaseous pollutants is too low to make flaring practical. Widely used in industry, typically, the thermal combustion system operates at high temperatures. Within the thermal incinerator, the contaminant airstream passes around or through a burner and into a refractory-line residence chamber where oxidation occurs (see Figure 16.12). Residence time is the amount of time the fuel mixture remains in the combustion chamber. Flue gas from a thermal incinerator (which

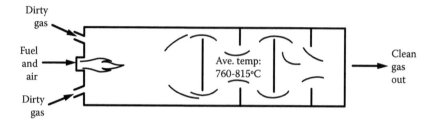

FIGURE 16.12 Thermal incinerator. From USEPA, *Control Techniques for Gases and Particulates*, 1968.

is relatively clean) is at high temperature and contains recoverable heat energy. Thermal incinerators can destroy gaseous pollutants at efficiencies of greater than 99% when operated correctly. Figure 16.13 shows a schematic of a typical thermal incinerator system.

Catalytic Combustion

Catalytic combustion operates by passing a preheated contaminant-laden gas stream through a catalyst bed (usually thinly coated platinum mesh mat, honeycomb, or other configuration designed to increase surface area), which promotes the oxidization reaction at lower temperatures (see Figure 16.14). The metal catalyst is used to initiate and promote combustion at much lower temperatures than those required for thermal combustion (metals in the platinum family are recognized for their ability to promote combustion at low temperature). Catalytic incineration may require 20 to 50 times less residence time than thermal incineration (see Table 16.4 for other advantages of catalytic incinerators over thermal incinerators). Catalytic incinerators normally operate at 700–900°F. At this reduced temperature range, a savings in fuel usage and cost is realized; however, this may be offset by the cost of the catalytic incinerator itself. Destruction efficiencies greater than 95% are possible using a catalytic incinerator. Higher efficiencies are possible if larger catalyst volumes or higher temperatures are used.

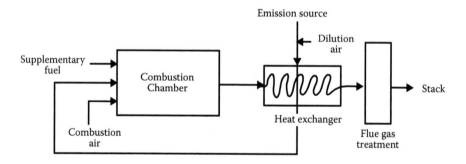

FIGURE 16.13 Schematic of thermal incinerator system. Adaptation from Corbitt (1990), p. 4.70.

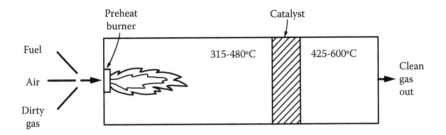

FIGURE 16.14 A catalytic incinerator. From USEPA, *Control Techniques for Gases and Particulates*, 1971.

A schematic diagram of a catalytic incinerator is presented in Figure 16.15. A heat exchanger is an option for systems with heat transfer between two gas streams (recuperative heat exchange). The need for dilution air, combustion air, and flue gas treatment is based on site-specific conditions. Catalysts are subject to both physical and chemical deterioration, and their usefulness is suppressed by sulfur-containing compounds. For best performance, catalyst surfaces must be clean and active.

Catalytic incineration is used in a variety of industries to treat effluent gases, including emissions from paint and enamel bake ovens, asphalt oxidation, coke ovens, formaldehyde manufacture, and varnish cooking. Catalytic incinerators are best suited for emission streams with low VOC content.

REMOVAL OF GASEOUS POLLUTANTS: MOBILE SOURCES

Mobile sources of gaseous pollutants include locomotives, ships, airplanes, and automobiles. However, automobiles are by far the most important in terms of both total emissions and location of emissions relative to people. To achieve on-road mobile source emission control, an integrated approach has been used. This integrated approach includes technological advances in vehicle and engine design together with clearers, high-quality fuels, plus the addition of vapor and particulate recovery systems and the development of auto inspection and maintenance (I/M) programs.

According to the *Twelfth Annual Report of the Council on Environmental Quality* (1982), transportation accounted for 55% of all major air pollutants emitted to the atmosphere in 1980. In 1986 there were almost 140 million registered passenger cars

---

TABLE 16.4
Advantages of Catalytic over Thermal Incinerators

1. Catalytic incinerators have lower fuel requirements.
2. Catalytic incinerators have lower operating temperatures.
3. Catalytic incinerators have little or no insulation requirements.
4. Catalytic incinerators have reduced fire hazards.
5. Catalytic incinerators have reduced flashback problems.

Source: Adapted from Buonicore and Davis (1992).

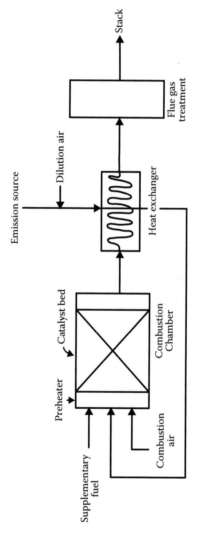

FIGURE 16.15 Schematic of catalytic incinerator system. Adaptation from Corbitt (1990), p. 4.71.

in the United States, consuming more than 1 billion gallons of fuel. Total emissions from these vehicles included 58% of the nation's total carbon monoxide, 38% of the lead, 34% of the nitrogen oxide, 27% of the VOCs, and 16% of the particulates (USEIA, 1988; USEPA, 1988).

Because of the high levels of pollutant emissions from the automobile, emission standards have become increasingly more stringent in the United States. Under the Clean Air Act of 1970, for example, motor vehicle emissions standards forced the development of new control technology to achieve compliance with the standard emission levels.

Mobile source pollution problems can be solved by either of two means: replacing the internal combustion engine (e.g., with electrical power or mass transit) or by direct pollutant control systems. At the present time replacement is not feasible and technology in this regard is still in the infancy stage. Even if the technology were currently available for replacement, replacement would be inordinately difficult to undertake. Direct pollutant control systems (those that control emissions from the crankcase, carburetor, fuel tank, and exhaust) are what we rely on. We discuss these systems in the following section.

CONTROL OF CRANKCASE EMISSIONS

Crankcase emissions can be controlled by technology called positive crankcase ventilation (PCV). In this control technology, hydrocarbon blowby gases (gases that go past the piston rings into the crankcase) are recirculated to the combustion chamber for reburning. The National Air Pollution Control Administration (1970) estimated that incorporation of the PCV system reduced crankcase hydrocarbon emissions to negligible levels.

CONTROL OF EVAPORATIVE EMISSIONS

Changes in ambient temperatures (diurnal losses), hot soak, and running losses result in evaporative emissions. Diurnal losses are caused by expansions of the air-fuel mixture in a partially filled fuel tank, which expels gasoline vapor into the atmosphere. Hot soak emissions occur after the engine is shut off as heat from the engine causes increased evaporation of fuel. Running (or operating) losses occur during driving as the fuel is heated by the road surface, and when fuel is forced from the fuel tank while the vehicle is being operated and the fuel tank becomes hot. In 1971 the direct control measure instituted to control hydrocarbon emissions was the installation of a canister filled with activated charcoal that adsorbs hydrocarbon emissions. Adsorbed vapors are purged from charcoal into the engine during high-power operating conditions. In California, vapor control systems at service stations reduce potential refueling vapor losses (Perkins, 1974).

CATALYTIC CONVERTERS

Beginning in 1975, all new U.S. automobiles were required to be equipped with catalytic converters to meet the more restrictive tailpipe emission standards. Three types of catalytic converters are used for this purpose: oxidizing, reducing, and three-way. Table 16.5 lists characteristics of each of these catalytic converters.

TABLE 16.5
Types of Catalytic Converters

Oxidizing catalytic converter	Works to accelerate the completion of the oxidation of CO and hydrocarbons so that CO is converted to carbon dioxide and water vapor. Platinum and palladium are used as catalysts. These converters can be poisoned by lead; thus, only unleaded gasoline should be used with this type.
Reducing catalytic converters	Uses rhodium and ruthenium to accelerate the reduction of NO_x to N_2.
Three-way catalytic converters	Favored by U.S. automobile manufacturers because it enables them to meet compliance requirements of the Clean Air Act. Oxidizes hydrocarbons and carbon monoxide to carbon dioxide, while reducing NO_x to N_2. Effective in controlling emissions and has the advantage of allowing the engine to operate at normal conditions where engine performance and efficiency are greatest.

Source: Adaptation from Demmler (1977).

AUTO INSPECTION AND MAINTENANCE (I/M) PROGRAMS

Under the 1990 Clean Air Act, auto manufacturers will build cleaner cars and cars will use cleaner fuels. However, to get air pollution down and keep it down, a third program is needed: vehicle inspection and maintenance (I/M), which makes sure cars are being maintained adequately to keep pollution emissions (releases) low. The 1990 Clean Air Act includes very specific requirements for inspection and maintenance programs.

REFERENCES AND MAINTENANCE (I/M) PROGRAMS

American Industrial Hygiene Association. 1968. *Air pollution manual: Control equipment, Part II.* Detroit: AIHA.

Boubel, R. W., Fox, D. L., Turner, D. B., and Stern, A. C. 1994. *Fundamentals of air pollution.* New York: Academic Press.

Buonicore, A. J., Theodore, L., and Davis, W. T. 1992. Air pollution control engineering. In Buonicore, A. J., and Davis, W. T., eds., *Air pollution engineering manual.* New York: Van Nostrand Reinhold.

Buonicore, A. J., and Davis, W. T., eds. 1992. *Air pollution engineering manual.* New York: Van Nostrand Reinhold.

Cooper, C. D., and Alley, F. C. 1990. *Air pollution control: A design approach.* Prospect Heights, IL: Waveland Press.

Corbitt, R. A. 1990. *Standard handbook of environmental engineering.* New York: McGraw-Hill.

Davis, M. L., and Cornwell, D. A. 1991. *Introduction to environmental engineering.* New York: McGraw-Hill.

Demmler, A. W. 1977. Automotive catalysis. *Auto Eng.* 85:29, 32.

DOE. *Twelfth annual report of the council on environmental quality.* 1982. Washington, DC.

Godish, T. 1997. *Air quality.* 3rd ed. Boca Raton, FL: Lewis Publishers.

Hesketh, H. E. 1991. *Air pollution control: Traditional and hazardous pollutants.* Lancaster, PA: Technomic Publishing Company.

Heumann, W. L. 1997. *Industrial air pollution control systems.* New York: McGraw-Hill.

National Air Pollution Control Administration. 1970. *Control techniques for hydrocarbon and organic solvent emissions for stationary sources.* Document B, publication AP-68. Washington, DC: Author.

Peavy, H. S., Rowe, D. R., and Tchobanglous, G. 1985. *Environmental engineering.* New York: McGraw-Hill.

Perkins, H. C. 1974. *Air pollution.* New York: McGraw-Hill.

Spellman, F. R., and Whiting, N. 2006. *Environmental science and technology: Concepts and applications.* 2nd ed. Rockville, MD: Government Institute.

USEIA. 1988. *Annual energy review 1987.* Washington, DC: Department of Energy, Author.

USEPA. 1971. *Annual report of the Environmental Protection Agency to the Congress of the United States in compliance with section 202(b)(4), Public Law 90-148.* Washington, DC: Author.

USEPA. 1973. *Air pollution engineering manual.* 2nd ed., AP-40, Research Triangle Park, NC: Author.

USEPA. 1986. *Handbook—Control technologies for hazardous air pollutants.* EPA 625/6-86/014. Cincinnati, Ohio: Center for Environmental Research Information, Author.

USEPA. 1988. *National air pollutant emission estimates 1940–1986.* Washington, DC: Author.

USEPA. 2006. Air pollution control orientation course. www//www.epa.gov/air/oaqps/eop/course422/ce6c.html (accessed January 26, 2008).

USHEW. 1969. *Control techniques for particulate air pollutants.* Washington, DC: National Air Pollution Control Administration.

17 Indoor Air Quality

The quality of the air we breathe and the attendant consequences for human health are influenced by a variety of factors. These include hazardous material discharges indoors and outdoors, meteorological and ventilation conditions, and pollutant decay and removal processes. Over 80% of our time is spent in indoor environments, so that the influence of building structures, surfaces, and ventilation are important considerations when evaluating air pollution exposures (Wadden and Scheff, 1983, p. 1).

INTRODUCTION

For those familiar with *Star Trek* (Trekees) and for those who are not, consider a quotable quote: "The air is the air." However, in regards to the air we breathe, according to USEPA (2001), few of us realize that we all face a variety of risks to our health as we go about our day-to-day lives. Driving our cars, flying in planes, engaging in recreational activities, and being exposed to environmental pollutants all pose varying degrees of risk. Some risks are simply unavoidable. Some we choose to accept because to do otherwise would restrict out ability to lead our lives the way we want. And some are risks we might decide to avoid if we had the opportunity to make informed choices. Indoor air pollution is one risk that we can do something about.

In the last several years, a growing body of scientific evidence has indicated that the air within homes and other buildings can be more seriously polluted than the outdoor air in even the largest and most industrialized cities. Other research indicates that people spend approximately 90% of their time indoors. Recall that in Chapter 11 we discussed microclimates. A type of microclimate we do not often think about (if at all) is the microclimates we spend 80% of our time in: the office or the home (indoors) (Wadden and Scheff, 1983). Thus, for many people, the risks to health may be greater due to exposure to air pollution indoors than outdoors (USEPA, 2001).

Not much attention was given to indoor microclimates until after two events took place a few years ago. The first event had to do with Legionnaires' disease, and the second with sick building syndrome. In addition, people who may be exposed to indoor air pollutants for the longest periods of time are often those most susceptible to the effects of indoor air pollution. Such groups include the young, the elderly, and the chronically ill, especially those suffering from respiratory or cardiovascular disease.

The impact of energy conservation on inside environments may be substantial, particularly with respect to decreases in ventilation rates (Hollowell et al., 1979a) and tight buildings constructed to minimize infiltration of outdoor air (Woods, 1980; Hollowell et al., 1979b). The purpose of constructing tight buildings is to save energy—to keep the heat or air conditioning inside the structure. The problem is indoor air contaminants within these tight structures are not only trapped within but also can be concentrated, exposing inhabitants to even more exposure.

These topics and others along with causal factors leading to indoor air pollution are covered in this chapter. What about indoor air quality problems in the workplace? In this chapter we also discuss this pervasive but often overlooked problem. In this regard, we discuss the basics of indoor air quality (IAQ; as related to the workplace environment) and the major contaminants that currently contribute to this problem. Moreover, mold and mold remediation, although not new to the workplace, are the new buzzwords attracting attention these days. Contaminants such as asbestos, silica, lead, and formaldehyde are also discussed. Various related remediation practices are also discussed.

LEGIONNAIRES' DISEASE

Since that infamous event that occurred in Philadelphia in 1976 at the Belleview Stratford Hotel during a convention of American Legion members, which included 182 cases and 29 deaths, *Legionella pneumophila* (the deadly bacterium) has become synonymous with the term *Legionnaires' disease*. The deaths were attributed to colonized bacteria in the air conditioning system cooling tower.

Let us take a look at this deadly killer—a killer that inhabits the microclimate we call office, hotel, and other indoor spaces.

Organisms of the genus *Legionella* are ubiquitous in the environment and are found in natural fresh water and potable water, as well as in closed-circuit systems, such as evaporative condensers, humidifiers, recreational whirlpools, air handling systems, and, of course, in cooling tower water.

The potential for the presence of *Legionella* bacteria is dependent on certain environmental factors: moisture, temperature (50–140°F), oxygen, and a source of nourishment such as slime or algae.

Not all the ways in which Legionnaires' disease can be spread are known to us at this time; however, we do know that it can be spread through the air. Centers for Disease Control (CDC) states (in its *Questions and Answers on Legionnaires' Disease*, CDC 28L0343779) that there is no evidence that Legionnaires' disease is spread person to person.

Air conditioning cooling towers and evaporative condensers have been the source of most outbreaks to date, and the bacterium is commonly found in both. Unfortunately, we do not know if this is an important means of spreading of Legionnaires' disease because other outbreaks have occurred in buildings that did not have air conditioning.

Not all people are at risk of contracting Legionnaires' disease. The people most at risk include persons:

1. With lowered immunological capacity
2. Who smoke cigarettes and abuse alcohol
3. Who are exposed to high concentrations of *Legionella pneumophila*

Most commonly recognized as a form of pneumonia, the symptoms of Legionnaires' disease usually become apparent 2–10 days after known or presumed exposure to airborne Legionnaires' disease bacteria. A sputum-free cough is common, but sputum production is sometimes associated with the disease. Within less than a day, the victim can experience rapidly rising fever and the onset of chills. Mental confusion, chest pain, abdominal pain, impaired kidney function, and diarrhea are associated manifestations of the disease. CDC estimates that around 25,000 people develop Legionnaires' disease annually.

The obvious question becomes: How do we prevent or control Legionnaires' disease? The controls presently being used are targeted on cooling towers and air handling units (condensate drain pans).

Cooling tower procedures used to control bacterial growth vary somewhat on the various regions in a cooling tower system. However, control procedures usually include a good maintenance program, including repair/replacement of damaged components, routine cleaning, and sterilization.

In sterilization, a typical protocol calls for the use of chlorine in a residual solution at about 50 ppm combined with a detergent that is compatible to produce the desired sterilization effect. It is important to ensure that even those spaces that are somewhat inaccessible are properly cleaned of slime and algae accumulations.

Regarding control measures for air handling units, condensate drain pans typically involve keeping the pans clean and checked for proper drainage of fluid. This is important to prevent stagnation and the buildup of slime/algae/bacteria. A cleaning and sterilization program is required anytime algae or slime are found in the unit.

SICK BUILDING SYNDROME

The second event that got the public's attention regarding microclimates and the possibility of unhealthy environments contained therein was actually spawned by the first, the Legionnaires' event and other incidents or complaints that followed. What we are referring to here is sick building syndrome.

The term *sick building syndrome* was coined by an international working group under the World Health Organization (WHO) in 1982. The WHO working group studied the literature about indoor climate problems and found that these microclimates in buildings are characterized by the same set of frequently appearing complaints and symptoms. WHO came up with five categories of symptoms exemplified by some complaints reported by occupants supposed to suffer from sick building syndrome (SBS). These categories are listed in the following:

1. *Sensory irritation in eyes, nose, and throat*—Pain, sensation of dryness, smarting feeling, stinging, irritation, hoarseness, voice problems.
2. *Neurological or general health symptoms*—Headache, sluggishness, mental fatigue, reduced memory, reduced capability to concentrate, dizziness, intoxication, nausea and vomiting, tiredness.
3. *Skin irritation*—Pain, reddening, smarting or itching sensations, dry skin.

4. *Nonspecific hypersensitivity reactions*—Running nose and eyes, asthma-like symptoms among nonasthmatics, sounds from the respiratory system.
5. *Odor and taste symptoms*—Changed sensitivity of olfactory or gustatory sense, unpleasant olfactory or gustatory perceptions.

In the past similar symptoms had been used to define other syndromes such as the building disease, the building illness syndrome, building-related illness, or the tight-fitting office syndrome, which in many cases appear to be synonyms for the sick building syndrome; thus, the WHO definition of the SBS worked to combine these syndromes into one general definition or summary. A summary compiled by WHO (1982, 1984) and Molhave (1986) of this combined definition includes the five categories of symptoms listed earlier and also:

1. Irritation of mucous membranes in eye, nose, and throat is among the most frequent symptoms.
2. Other symptoms, e.g., from lower airways or from internal organs, should be infrequent.
3. A large majority of occupants report symptoms.
4. The symptoms appear especially frequent in one building or in part of it.
5. No evident causality can be identified in relation either to exposures or to occupant sensitivity.

The WHO group suggested the possibility that the SBS symptoms have a common causality and mechanism (WHO, 1982). However, the existence of SBS is still a postulate because the descriptions of the symptoms in the literature are anecdotal and unsystematic (Molhave, 1986).

INDOOR AIR POLLUTION

Why is indoor air pollution a problem? As indicated above, recognition that the indoor air environment may be a health problem is a relatively recent emergence. The most significant problems of indoor air quality are the impact of cigarette smoking, stove and oven operation, and emanations from certain types of particleboard, cement, and other building materials (Wadden and Scheff, 1983).

The significance of the indoor air quality problem became apparent not only because of the Legionnaires' incident of 1976 and the WHO study of 1982, but also because of another factor that came to the forefront in the mid-1970s: the need to conserve energy. In the early 1970s, when hundreds of thousands of people were standing in line to obtain gasoline for their automobiles, it was not difficult to drive home the need to conserve energy supplies.

The resulting impact of energy conservation on inside environments has been substantial. This is especially the case in regards to building modifications that were made to decrease ventilation rates and new construction practices that were incorporated to ensure tight buildings to minimize infiltration of outdoor air.

There is some irony in this development, of course. While there is a need to ensure proper building design, construction, and ventilation guidelines to avoid the exposure of inhabitants to unhealthy environments, what really resulted in this mad dash to reduce ventilation rates and tighten buildings from infiltration was a trade-off: energy economics versus air quality.

According to Byrd (2003), indoor air quality (IAQ) refers to the effect, good or bad, of the contents of the air inside a structure on its occupants. Stated differently, IAQ, in this text, refers to the quality of the air inside workplaces as represented by concentrations of pollutants and thermal (temperature and relative humidity) conditions that affect the health, comfort, and performance of employees. Usually, temperature (too hot or too cold), humidity (too dry or too damp), and air velocity (draftiness or motionlessness) are considered comfort rather than indoor air quality issues. Unless they are extreme, they may make someone uncomfortable, but they will not make a person ill. Other factors affecting employees, such as light and noise, are important indoor environmental quality considerations, but are not treated as core elements of indoor air quality problems. Nevertheless, most industrial hygienists must take these factors into account in investigating environmental quality situations.

Byrd (2003) further points out that "good IAQ is the quality of air, which has no unwanted gases or particles in it at concentrations, which will adversely affect someone. Poor IAQ occurs when gases or particles are present at an excessive concentration so as to affect the satisfaction of health of occupants."

In the workplace, poor IAQ may only be annoying to one person; however, at the extreme, it could be fatal to all the occupants in the workplace.

The concentration of the contaminant is crucial. Potentially infectious, toxic, allergenic, or irritating substances are always present in the air. Note that there is nearly always a threshold level below which no effect occurs.

COMMON INDOOR AIR POLLUTANTS IN THE HOME

This section takes a brief source-by-source look at the most common indoor air pollutants, their potential health effects, and ways to reduce their levels in the home.

RADON

Radon is a noble, nontoxic, colorless, odorless gas produced in the decay of radium-226 and is found everywhere at very low levels. Radon is ubiquitously present in the soil and air near to the surface of the earth. As radon undergoes radioactive decay, it releases an alpha particle, gamma ray, and progeny that quickly decay to release alpha and beta particles and gamma rays. Because radon progeny are electrically charged, they readily attach to particles, producing a radioactive aerosol. It is when radon becomes trapped in buildings and concentrations build up in indoor air that exposure to radon becomes of concern. This is the case because aerosol radon-contaminated particles may be inhaled and deposited in the bifurcations of respiratory airways. Irradiation of tissue at these sites poses a significant risk of lung cancer (depending on exposure dose).

How does radon enter a house? The most common way in which radon enters a house is through the soil or rock upon which the house is built. The most common source of indoor radon is uranium, which is common to many soils and rocks.

As uranium breaks down, it releases soil or radon gas, and radon gas breaks down into radon decay products or progeny (commonly called radon daughters). Radon gas is transported into buildings by pressure-induced convective flows.

There are other sources of radon, for example, from well water and masonry materials.

Radon levels in a house vary in response to temperature-dependent and wind-dependent pressure differentials and to changes in barometric pressures. When the base of a house is under significant negative pressure, radon transport is enhanced.

Studies by the USEPA (1987, 1988) indicate that as many as 10% of all American homes, or about 9 million homes, may have elevated levels of radon, and the percentage may be higher in geographic areas with certain soils and bedrock formations.

According to USEPA booklets *A Citizen's Guide to Radon*, *Radon Reduction Methods: A Homeowner's Guide*, and *Radon Measurement Proficiency Report* (for each state), exposure to radon in the home can be reduced by the following steps:

1. Measure levels of radon in the home.
2. The state radiation protection office can provide you with information on the availability of detection devices or services.
3. Refer to EPA guidelines in deciding whether and how quickly to take action based on test results.
4. Learn about control measures.
5. Take precautions not to draw larger amounts of radon into the house.
6. Select a qualified contractor to draw up and implement a radon mitigation plan.
7. Stop smoking and discourage smoking in your home.
8. Treat radon-contaminated well water by aerating or filtering through granulated activated charcoal.

ENVIRONMENTAL TOBACCO SMOKE

The use of tobacco products by approximately 45 million smokers in the United States results in significant indoor contamination from combustion by-products that pose significant exposures to millions of others who do not smoke but who must breathe contaminated indoor air. Composed of sidestream smoke (smoke that comes from the burning end of a cigarette) and smoke that is exhaled by the smoker, tobacco contains a complex mixture of over 4,700 compounds, including both gases and particles.

According to reports issued in 1986 by the surgeon general and the National Academy of Sciences, environmental tobacco smoke is a cause of disease, including lung cancer, in both smokers and healthy nonsmokers. Environmental tobacco smoke may also increase the lung cancer risk associated with exposures to radon.

The following steps can reduce exposure to environmental tobacco smoke in the office and home:

1. Give up smoking and discourage smoking in your home and place of work, or require smokers to smoke outdoors.
2. A common method of reducing exposure to indoor air pollutants such as environmental tobacco smoke is ventilation that works to reduce but not eliminate exposure.

Biological Contaminants

A variety of biological contaminants can cause significant illness and health risks. These include mold and mildew, viruses, animal dander and cat saliva, mites, cockroaches, pollen, and infections from airborne exposures to viruses that cause colds and influenza and bacteria that cause Legionnaires' disease and tuberculosis (TB).

The following steps can reduce exposure to biological contaminants in the home and office:

1. Install and use exhaust fans that are vented to the outdoors in kitchens and bathrooms, and vent clothes dryers outdoors.
2. Ventilate the attic and crawl spaces to prevent moisture buildup.
3. Keep water trays in cool mist or ultrasonic humidifiers clean and filled with fresh distilled water daily.
4. Water-damaged carpets and buildings materials should be thoroughly dried and cleaned within 24 hours.
5. Maintain good housekeeping practices in both the home and office.

Combustion By-Products

Combustion by-products are released into indoor air from a variety of sources. These include unvented kerosene and gas space heaters, woodstoves, fireplaces, gas stoves, and hot water heaters. The major pollutants released from these sources are carbon monoxide, nitrogen dioxide, and particles.

The following steps can reduce exposure to combustion products in the home and office:

1. Fuel-burning unvented space heaters should only be operated using great care and special safety precautions.
2. Install and use exhaust fans over gas cooking stoves and ranges and keep the burners properly adjusted.
3. Furnaces, flues, and chimneys should be inspected annually, and any needed repairs should be made promptly.
4. Woodstove emissions should be kept to a minimum.

Household Products

A large variety of organic compounds are widely used in household products because of their useful characteristics, such as the ability to dissolve substances and evaporate quickly. Cleaning, disinfecting, cosmetic, degreasing, and hobby products all contain organic solvents, as do paints, varnishes, and waxes. All of these products can release organic compounds while using them and when they are stored.

The following steps can reduce exposure to household organic compounds:

1. Always follow label instructions carefully.
2. Throw away partially full containers of chemicals safely.
3. Limit the amount you buy.

PESTICIDES

Pesticides represent a special case of chemical contamination of buildings, where the EPA estimates 80–90% of most people's exposure in the air occurs. These products are extremely dangerous if not used properly.

The following steps can reduce exposure to pesticides in the home:

1. Read the label and follow directions.
2. Use pesticides only in well-ventilated areas.
3. Dispose of unwanted pesticides safely.

ASBESTOS IN THE HOME

Asbestos became a major indoor air quality concern in the United States in the late 1970s. Asbestos is a mineral fiber commonly used in a variety of building materials and has been identified as having the potential (when friable) to cause cancer in humans.

The following steps can reduce exposure to asbestos in the home and office:

1. Do not cut, rip, or sand asbestos-containing materials.
2. When you need to remove or clean up asbestos, use a professional, trained contractor.

WHY IS IAQ IMPORTANT TO WORKPLACE OWNERS?

Workplace structures (buildings) exist to protect workers from the elements and to otherwise support worker activity. Workplace buildings should not make workers sick, cause them discomfort, or otherwise inhibit their ability to perform. How effectively a workplace building functions to support its workers and how efficiently the workplace building operates to keep costs manageable is a measure of the workplace building's performance.

The growing proliferation of chemical pollutants in industrial and consumer products, the tendency toward tighter building envelopes and reduced ventilation to save energy, and pressures to defer maintenance and other building services to reduce costs have fostered indoor air quality problems in many workplace buildings. Employee complaints of odors, stale and stuffy air, and symptoms of illness or discomfort breed undesirable conflicts between workplace occupants and workplace managers. Lawsuits sometimes follow.

If indoor air quality is not well managed on a daily basis, remediation of ensuing problems or resolution in court can be extremely costly. Moreover, air quality problems in the workplace can lead to reduced worker performance. So it helps to understand the causes and consequences of indoor air quality and to manage your workplace buildings to avoid these problems.

WORKER SYMPTOMS ASSOCIATED WITH POOR AIR QUALITY

Worker responses to pollutants, climatic factors, and other stressors such as noise and light are generally categorized according to the type and degree of responses and the timeframe in which they occur. Workplace managers should be generally familiar with these categories, leaving detailed knowledge to industrial hygienists.

- *Acute effects*—Acute effects are those that occur immediately (e.g., within 24 hours) after exposure. Chemicals released from building materials may cause headaches, or mold spores may result in itchy eyes and runny nose in sensitive individuals shortly after exposure. Generally, these effects are not long lasting and disappear shortly after exposure ends. However, exposure to some biocontaminants (fungi, bacteria, viruses) resulting from moisture problems, poor maintenance, or inadequate ventilation has been known to cause serious, sometimes life-threatening respiratory diseases that themselves can lead to chronic respiratory conditions.
- *Chronic effects*—Chronic effects are long-lasting responses to long-term or frequently repeated exposures. Long-term exposures to even low concentrations of some chemicals may induce chronic effects. Cancer is the most commonly associated long-term health consequence of exposure to indoor air contaminants. For example, long-term exposures to environmental tobacco smoke, radon, asbestos, and benzene increase cancer risk.
- *Discomfort*—Discomfort is typically associated with climatic conditions, but workplace building contaminants may also be implicated. Workers complain of being too hot or too cold or experience eye, nose, or throat irritation because of low humidity. However, reported symptoms can be difficult to interpret. Complaints that the air is "too dry" may result from irritation from particles on the mucous membranes rather than low humidity, or "stuffy air" may mean that the temperature is too warm or there is lack of air movement, or "stale air" may mean that there is a mild but difficult to identify odor. These conditions may be unpleasant and cause discomfort among workers, but there is usually no serious health implication involved. Absenteeism, work performance, and employee morale, however, can be seriously affected when building managers fail to resolve these complaints.
- *Performance effects*—Significant measurable changes in workers' ability to concentrate or perform mental or physical tasks have been shown to result from modest changes in temperature and relative humidity. In addition, recent studies suggest that similar effects are associated with indoor pollution due to lack of ventilation or the presence of pollution sources. Estimates of performance losses from poor indoor air quality for all buildings suggest a 2–4% loss on average. Future research should further document and quantify these effects.

BUILDING FACTORS AFFECTING INDOOR AIR QUALITY

Building factors affecting indoor air quality can be grouped into factors affecting indoor climate and factors affecting indoor air pollution.

- *Factors affecting indoor climate*—The thermal environment (temperature, relative humidity, and airflow) are important dimensions of indoor air quality for several reasons. First, many complaints of poor indoor air may be resolved by simply altering the temperature or relative humidity. Second, people that are thermally uncomfortable will have a lower tolerance to other building discomforts. Third, the rate at which chemicals are released from building material is usually higher at higher building temperatures. Thus, if occupants are too warm, it is also likely that they are being exposed to higher pollutant levels.

- *Factors affecting indoor air pollution*—Much of the building fabric, its furnishings and equipment, its occupants and their activities produce pollution. In a well-functioning building, some of these pollutants will be directly exhausted to the outdoors and some will be removed as outdoor air enters that building and replaces the air inside. The air outside may also contain contaminants, which will be brought inside in this process. This air exchange is brought about by the mechanical introduction of outdoor air (outdoor air ventilation rate), the mechanical exhaust of indoor air, and the air exchanged through the building envelope (infiltration and exfiltration).

Pollutants inside can travel through the building as air flows from areas of higher atmospheric pressure to areas of lower atmospheric pressure. Some of these pathways are planned and deliberate so as to draw pollutants away from occupants, but problems arise when unintended flows draw contaminants into occupied areas. In addition, some contaminants may be removed from the air through natural processes, as with the adsorption of chemicals by surfaces or the settling of particles onto surfaces. Removal processes may also be deliberately incorporated into the building systems. Air filtration devices, for example, are commonly incorporated into building ventilation systems.

Thus, the factors most important to understanding indoor pollution are (1) indoor sources of pollution, (2) outdoor sources of pollution, (3) ventilation parameters, (4) airflow patterns and pressure relationships, and (5) air filtration systems.

TYPES OF WORKPLACE AIR POLLUTANTS

Common pollutants or pollutant classes of concern in commercial buildings along with common sources of these pollutants are provided in Table 17.1.

SOURCES OF WORKPLACE AIR POLLUTANTS

Air quality is affected by the presence of various types of contaminants in the air. Some are in the form of gases. These would be generally classified as toxic chemicals. The types of interest are combustion products (carbon monoxide, nitrogen dioxide), volatile organic compounds (formaldehyde, solvents, perfumes and fragrances, etc.), and semivolatile organic compounds (pesticides). Other pollutants are in the form of animal dander, etc.; soot; particles from buildings, furnishings, and occupants such as fiberglass, gypsum powder, paper dust, lint from clothing, carpet fibers, etc.; dirt (sandy and earthy material), etc.

Burge and Hoyer (1998) point out many specific sources for contaminants that result in adverse health effects in the workplace, including the workers (contagious

TABLE 17.1
Indoor Pollutants and Potential Sources

Pollutant or Pollutant Class	Potential Sources
Environmental tobacco smoke	Lighted cigarettes, cigars, pipes
Combustion contaminants	Furnaces, generators, gas or kerosene space heaters, tobacco products, outdoor air, vehicles
Biological contaminants	Wet or damp materials, cooling towers, humidifiers, cooling coils or drain pans, damp duct insulation or filters, condensation, reentrained sanitary exhausts, bird droppings, cockroaches or rodents, dust mites on upholstered furniture or carpeting, body odors
Volatile organic compounds (VOCs)	Paints, stains, varnishes, solvents, pesticides, adhesives, wood preservatives, waxes, polishes, cleansers, lubricants, sealants, dyes, air fresheners, fuels, plastics, copy machines, printers, tobacco products, perfumes, dry cleaned clothing
Formaldehyde	Particleboard, plywood, cabinetry, furniture, fabrics
Soil gases (radon, sewer gas, VOCs, drain leak, dry drain methane)	Soil and rock (radon), sewer traps, leaking underground storage tanks, landfill
Pesticides	Termiticides, insecticides, rodenticides, fungicides, disinfectants, herbicides

diseases, carriage of allergens and other agents on clothing); building compounds (VOCs, particles, fibers); contamination of building components (allergens, microbial agents, pesticides); and outdoor air (microorganisms, allergens, chemical air pollutants).

When workers complain of IAQ problems, the industrial hygienist is called upon to determine if the problem really is an IAQ problem. If he or she determines that some form of contaminant is present in the workplace, proper remedial action is required. This usually includes removing the source of the contamination.

THE CASE OF THE STICKY HEAD

In 1996 (on a daily basis), a supervisor complained about a sticky, perfume-laden, gooey, messy substance that accumulated on his bald head every time he sat at his desk in his second-floor office. Wondering what was causing this unusual occurrence, the supervisor finally reported the mysterious daily accumulation of goo to the organizational safety professional.

The organizational safety professional, a certified industrial hygienist (CIH), not only saw the humor in the supervisor's goo report but also understood the more serious implication: something was afoul (in more than one way) with second-floor IAQ. In particular, something was peculiar, strange, and not right about the operation of the second floor's HVAC system.

The safety professional quickly identified the perfume-laden goo delivery vehicle: the very large-diameter ventilation supply diffuser—located directly above the bald-headed supervisor's desk chair. The safety professional attached a plain piece of white copy paper that hung a few inches below the diffuser's pin-holed-sized outlets and waited.

After a 2-hour wait, the safety professional removed the paper from the diffuser. She discovered that what she was now holding in her hand was no longer a plain, white piece of copy paper, but instead a perfect sheet of yellowish, sticky flypaper. Additionally, there was no need to hold the sticky paper close to her breathing zone to notice the heavy, sweet smell of perfumed hairspray oozing, wafting from it.

So, with the problem in hand, so to speak, the safety professional then had to determine why women's hairspray, perfume, and whatever else was coming out of the ventilation supply diffuser right above the supervisor's desk chair. The air supply should have been from the building's makeup air supply, which comes from outside the building, but obviously it was coming from another source—the wrong source.

The safety supervisor began her investigation of the perfumed air source by tracing the overhead ventilation ducting hand over hand from room to room. Approximately 100 feet along the overhead ducting from the bald-headed supervisor's office the safety professional noticed that all the ductwork came together into a central, square-boxed metal unit. She noticed that there were two of these central, square-boxed ducts; they were labeled "supply" and "exhaust."

Standing on an 8-foot stepladder and craning her neck ceiling-ward, the safety professional was at first confused by the octopi-like ductwork entering and leaving the octopi-like body of supply and exhaust boxes, which acted like distribution plenum areas for moving air into (supply) or out of (exhaust) the system. Trying to follow the maze of ducting to and from these distribution boxes was not easy. However, one thing she noticed right away: the supply ductwork from the bald-headed supervisor's office was not connected to the distribution box that was stenciled "supply" but instead was attached to the box stenciled "exhaust." She quickly realized that the supervisor's line was incorrectly attached to the wrong distribution box. To fix the problem seemed simple enough: disconnect the supply duct from the exhaust distribution plenum and connect it to the supply distribution box as it should be.

The fix was not as simple as she first imagined. After thinking about it a few minutes, she realized that there was another problem. Where was all the perfumed, sticky air coming from? To answer this problem, she traced each of the input side ductwork pipes connected to the distribution box back to the point where their exhaust diffusers were located to pick up air that was to be exhausted.

During the ductwork tracing operation, the safety supervisor discovered other problems—all the fire/smoke damper lever arms installed at connections where ducts formed junctions with other ducts were in the wrong position. The dampers (located inside the ductwork) direct airflow where wanted or prevent it from moving where not wanted (as fire/smoke dampers, they protect ductwork penetrations in walls or floors that have a fire resistance rating and perform operational smoke control in static or dynamic smoke management systems). With the dampers in the wrong positions, airflow was being directed where it was not supposed to flow, or there was no flow at all in some ducts.

Even though she identified more than a dozen dampers that were positioned in the improper position, the safety professional left the dampers in the position in which she found them (she did not want to change anything that would prevent her from understanding the problem). She continued her hand-over-hand duct tracing. This process took a few hours but finally led to her gaining a full understanding of the problem. At the terminal endpoints of the ducts she found 24 ceiling intake exhaust diffusers positioned above a large second-floor area that had been partitioned into 60 separate office spaces. The office spaces were occupied by 60 administrative clerks. All of the clerks were female.

The safety supervisor now had a pretty good inclination as to the source of the sticky hairspray and perfume that the bald-headed supervisor had complained about. However, as a professional she wanted to actually observe what she assumed to be the case—to be sure of her assumptions. Thus, she spent the next full workday walking through the partitioned office area at various times and observed the clerks within their partitioned work areas. She noticed that at lunchtime, many of the woman used hairspray to groom their hair and several others applied perfume from spray bottles. She also noticed that just a few minutes before completing the workday and leaving the building for home, many of the clerks performed the same routine of hairspray and perfume application.

The safety professional observed a couple of these hairspray applications and noticed that she could actually see some of the aerosol from spray cans directed at each head of hair. She also noticed that some of the aerosol was pulled upward, toward the nearest exhaust diffuser. Once inside the diffuser, the aerosol-laden air flowed directly to the distribution box area, where it should have been pulled into the main exhaust duct to the exterior of the building. Instead, when the aerosol-laden airstream left the distribution box, it was pushed into the wrongly connected air supply duct that ventilated the bald-headed supervisor's office.

The safety professional hired a ventilation contractor who specialized in correcting HVAC problems. The contractor was tasked with a threefold project: remove the sticky material (hairspray and perfume residue) coating the interior exhaust ducting; reconnect correctly the proper ductwork to its proper distribution box; and place each smoke/fire damper in the correct position. While placing the dampers in their designed positions, the safety professional had the contractor fasten and lock each mechanical damper (those that had to be manipulated by hand to change position) so that no one could intentionally and incorrectly reposition the damper without unlocking them. The automatic smoke and fire dampers were left unlocked to close in case of fire or smoke.

After the contractor completed the work, the safety professional assembled a team of observers and placed an observer at an exterior exhaust outlet duct, several observers at various supply and exhaust ducts, and she took up station right beneath the bald-headed supervisor's supply diffuser. Each observer (equipped with walkie talkie) was ordered to direct a spray can of observation tracer smoke into exhaust diffusers while the other observers simply watched to ensure that none of the smoke came out of the wrong diffuser. The safety professional watched to make sure that the bald-headed supervisor's air supply diffuser continued to supply conditioned air without any trace of smoke—conditioned fresh air.

After four separate successful tests in a row, the safety supervisor directed the observers to leave their posts and assemble in the conference room. In the conference room, the safety observer listened as each observer gave his or her report of what he or she had observed. Satisfied with the results of the study, the safety observer thanked and dismissed the observers. Later she reported to the bald-headed supervisor that he no longer had to fear a sticky head while working in his office. A week later, the safety professional conducted an employee meeting and stressed the importance of the building's HVAC system. She also explained that the building's HVAC system was not to be tampered with or adjusted by anyone without her permission.

Tables 17.2 and 17.3 identify indoor and outdoor sources, respectively, of contaminants commonly found in the workplace and offer some measures for maintaining control of these contaminants.

TABLE 17.2
Indoor Sources of Contaminants

Category/Common Sources	Mitigation and Control
Housekeeping and Maintenance	
Cleanser	Use low-emitting products
Waxes and polishes	Avoid aerosols and sprays
Disinfectants	Dilute to proper strength
Air fresheners	Do not overuse; use during unoccupied hours
Adhesives	Use proper protocol when diluting and mixing
Janitor's/storage closets	Store properly with containers closed and lid tight
Wet mops	Use exhaust ventilation for storage spaces (eliminate return air)
Drain cleaners	Clean mops, store mop top up to dry
Vacuuming	Avoid "air fresheners"—clean and exhaust instead
Paints and coatings	Use high-efficiency vacuum bags/filters
Solvents	Use integrated pest management
Pesticides	
Lubricants	
Occupant-Related Sources	
Tobacco products	Smoking policy
Office equipment (printers/copiers)	Use exhaust ventilation with pressure control for major local sources
Cooking/microwave	Low-emitting art supplies/marking pens
Art supplies	Avoid paper clutter
Marking pens	Education material for occupants and staff
Paper products	
Personal products (e.g., perfume)	
Tracked-in dirt/pollen	
Building Uses as Major Sources	
Print/photocopy shop	Use exhaust ventilation and pressure control
Dry cleaning	Use exhaust hoods where appropriate; check hood airflows
Science laboratory	
Medical office	
Hair/nail salon	
Cafeteria	
Pet store	
Building-Related Sources	
Plywood/compressed wood	Use low-emitting sources
Construction adhesives	Air out in an open/ventilated area before installing
Asbestos products	Increase ventilation rates during and after installing
Insulation	Keep material dry prior to enclosing
Wall/floor coverings (vinyl/plastic)	
Carpets/carpet adhesives	
Wet building products	

TABLE 17.2
(Cont.)

Category/Common Sources	Mitigation and Control
Building-Related Sources	

Category/Common Sources	Mitigation and Control
Transformers	
Upholstered furniture	
Renovation/remodeling	
HVAC System	
Contaminated filters	Perform HVAC preventive maintenance
Contaminated duct lining	Change filter
Dirty drain pans	Clean drain pans; proper slope and drainage
Humidifiers	Use portable water for humidification
Lubricants	Keep duct lining dry; move lining outside of duct if possible
Refrigerants	Fix leaks/clean spills
Mechanical room	Maintain spotless mechanical room (not a storage area)
Maintenance activities	Avoid back drafting
Combustion appliances (boilers/furnaces/stoves/generators)	Check/maintain flues from boiler to outside
	Keep combustion appliances properly tuned
	Disallow unvented combustion appliances
	Perform polluting activities during unoccupied hours
Moisture	
Mold	Keep building dry
Vehicles	
Underground/attached garage	Use exhaust ventilation
	Maintain garage under negative pressure relative to the building
	Check airflow patterns frequently
	Monitor CO

Table 17.3 identifies common sources of contaminants that are introduced from outside buildings. These contaminants frequently find their way inside through the building shell, openings, or other pathways to the inside.

INDOOR AIR CONTAMINANT TRANSPORT

Air contaminants reach worker breathing zones by traveling from the source to the worker by various pathways. Normally, the contaminants travel with the flow of air. Air moves from areas of high pressure to areas of low pressure. That is why controlling workplace air pressure is an integral part of controlling pollution and enhancing building IAQ performance.

Air movements should be from occupants, toward a source, and out of the building rather than from the source to the occupants and out the building. Pressure differences will control the direction of air motion and the extent of occupant exposure.

TABLE 17.3
Outdoor Sources of Contaminants

Category/Common Sources	Mitigation and Control
Ambient Outdoor Air	
Air quality in the general area	Filtration or air cleaning of intake air
Vehicular Sources	
Local vehicular traffic	Locate air intake away from source
Vehicle idling areas	Require engines shut off at loading dock
Loading dock	Pressurize building/zone
	Add vestibules/sealed doors near source
Commercial/Manufacturing Sources	
Laundry or dry cleaning	Locate air intake away from source
Paint shop	Pressurize building relative to outdoors
Restaurant	Consider air cleaning options for outdoor air intake
Photoprocessing	Use landscaping to block or redirect flow of contaminants
Automotive shop/gas station	
Electronics manufacture/assembly	
Various industrial operations	
Utilities/Public Works	
Utility power plant	
Incinerator	
Water treatment plant	
Agricultural	
Pesticide spraying	
Processing or packing plants	
Ponds	
Construction/Demolitions	
Pressurize building	
Use walk-off mats	
Building Exhaust	
Bathroom exhaust	Separate exhaust or relief from air intake
Restaurant exhaust	Pressurize building
Air handler relief vent	
Exhaust from major tenant (e.g., dry cleaner)	
Water Sources	
Pools of water on roof	Proper roof drainage
Cooling tower mist	Separate air intake from source of water
	Treat and maintain cooling tower water
Birds and Rodents	
Fecal contaminants	Bird proof intake grills
Bird nesting	Consider vertical grills

TABLE 17.3
(Cont.)

Category/Common Sources	Mitigation and Control
Building Operations and Maintenance	
Trash and refuse area	Separate source from air intake
Chemical/fertilizer/groundskeeping storage	Keep source area clean/lids on tight
Painting/roofing/sanding	Isolate storage area from occupied areas
Ground Sources	
Soil gas	Depressurize soil
Sewer gas	Seal foundation and penetrations to foundations
Underground fuel storage tanks	Keep air ducts away from ground sources

Driving forces change pressure relationships and create airflow. Common driving forces are identified in Table 17.4.

INDOOR AIR DISPERSION PARAMETERS

Several parameters (some characterize observed patterns of contaminant distribution and others characterize flow features, such as stability) are used to characterize the dispersion of a contaminant inside a room. Table 17.5 lists these parameters and groups them under four headings: contaminant distribution, temperature distribution, stability and buoyancy of the room air, and supply air conditions. Because of overlap between these subjects, some parameters appear more than once.

TABLE 17.4
Major Driving Forces

Driving Force	Effect
Wind	Positive pressure is created on the windward side, causing infiltration, and negative pressure on the leeward side, causing exfiltration, though wind direction can be varied due to surrounding structures.
Stack effect	When the air inside is warmer than outside, it rises, sometimes creating a column of rising—up stairwells, elevator shafts, vertical pipe chases, etc. This buoyant force of the air results in positive pressure on the higher floors, negative pressure on the lower floors, and a neutral pressure plane somewhere between.
HVAC/fans	Fans are designed to push air in a directional flow and create positive pressure in front and negative pressure behind the fan.
Flues and exhaust	Exhausted air from a building will reduce the building air pressure relative to the outdoors. Air exhausted will be replaced either through infiltration or through a planned outdoor air intake vent.
Elevators	The pumping action of a moving elevator can push air out of or draw air into the elevator shaft as it moves.

TABLE 17.5
Parameters Used to Characterize Contaminant Dispersion in Rooms

Contaminant Distribution	Temperature Distribution	Stability/Buoyancy Room Air	Supply Air Conditions
Contaminant concentration	Air diffusion performance index	Reynolds number	Purging effectiveness of inlet
Local mean age of air	Temperature effectiveness	Rayleigh number	Reynolds number
Purging effectiveness of inlets	Effective draft temperature	Grashof number	Froude number
Local specific contaminant-accumulating index		Froude number	Archimedes number
Air change efficiency		Richardson number	
Ventilation effectiveness factor		Flux Richardson number	
Relative ventilation efficiency		Buoyancy flux	

Source: ASHARAE, *Handbook: HVAC Systems and Equipment* (2000), *Fundamentals* (2001), *Refrigeration* (2002), and *HVAC Applications* (2003).

Did You Know?

Flow parameters such as the mean age of air are difficult but not impossible to calculate experimentally. They are used mainly as a tool to help interpret data from numerical simulations of contaminant dispersion (Peng and Davidson, 1999).

- *Contaminant concentration*—Indicator of contaminant distribution in a room, i.e., the mass of contaminant per unit volume of air (measured in kg/m^3).
- *Local mean age of air*—Average time it takes for air to travel from the inlet to any point *P* in the room (Di Tommaso et al., 1999).
- *Purging effectiveness of inlets*—Quantity that can be used to identify the relative performance of each inlet in a room where there are multiple inlets.
- *Local specific contaminant-accumulating index*—General index capable of reflecting the interaction between the ventilation flow and a specific contaminant source.
- *Air change efficiency* (ACE)—Measure of how effectively the air present in a room is replaced by fresh air from the ventilation system (Di Tommaso et al., 1999). It is the ratio of room mean age that would exist if the air in the room were completely mixed to the average time of replacement of the room.
- *Ventilation effectiveness factor* (VEF)—Ratio of two contaminant concentration differentials: the contaminant concentration in the supply air

(typically zero) and the contaminant concentration in the room under complete mixing conditions (Zhang et al., 2001).

- *Relative ventilation efficiency*—Ratio of the local mean age that would exist if the air in the room were completely mixed to the local mean age that is actually measured at a point.
- *Air diffusion performance index* (ADPI)—Primarily a measure of occupant comfort rather than an indicator of contaminant concentrations. It expresses the percentage of locations in an occupied zone that meet air movement and temperature specifications for comfort.
- *Temperature effectiveness*—Similar in concept to ventilation effectiveness and reflects the ability of a ventilation system to remove heat.
- *Effective draft temperature*—Indicates the feeling of coolness due to air motion.
- *Reynolds number*—Expresses the ratio of the inertial forces to viscous forces.
- *Rayleigh number*—Characterizes natural convection flows.
- *Grashof number*—Equivalent to the Rayleigh number divided by the Prandtl number (a dimensionless number approximating the ratio of viscosity and thermal diffusity).
- *Froude number*—Dimensionless number used to characterize flow through corridors and doorways and in combined displacement and wind ventilation cases.
- *Richardson number*—Characterizes the importance of buoyancy.
- *Flux Richardson number*—Used to characterize the stabilizing effect of stratification on turbulence.
- *Buoyancy flux*—Used to characterize buoyancy-driven flows.
- *Archimedes number*—Conditions of the supplied air are often characterized by the discharge Archimedes number, which expresses the ratio of the buoyancy forces to momentum forces or the strength of natural convection to forced convection.

COMMON AIRFLOW PATHWAYS

Contaminants travel along pathways—sometimes over great distances. Pathways may lead from an indoor source to an indoor location or from an outdoor source to an indoor location.

The location experiencing a pollution problem may be close by, in the same area, or in an adjacent area, but it may be a great distance from a contaminant source or on a different floor from the source.

Knowledge of common pathways helps to track down the source and prevent contaminants from reaching building occupants (see Table 17.6).

MAJOR IAQ CONTAMINANTS

Industrial hygienists spend a large portion of their time working with and mitigating air contaminant problems in the workplace. The list of potential contaminants workers might be exposed to while working is extensive. There are, however, a few major chemical- and material-derived air contaminants (other than those poisonous

TABLE 17.6
Common Airflow Pathways for Contaminants

Common Pathway	Comment
Indoors	
Stairwell/elevator shaft	The stack effect brings about airflow by drawing air toward these chases on the lower floors and elevator shaft, away from these chases on the higher floors, affecting the flow of contaminants.
Vertical electrical or plumbing chases	
Receptacles, outlets, openings	Contaminants can easily enter and exit building cavities and thereby move from space to space.
Duct or plenum	Contaminants are commonly carried by the HVAC system throughout the occupied spaces.
Duct or plenum leakage	Duct leakage accounts for significant unplanned airflow and energy loss in buildings.
Flue or exhaust leakage	Leaks from sanitary exhausts or combustion flues can cause serious health problems.
Room spaces	Air and contaminants move within a room or through doors and corridors to adjoining spaces.
Outdoors to Indoors	
Indoor air intake	Polluted outdoor air or exhaust air can enter the building through the air intake.
Windows/doors	A negatively pressurized building will draw air and outside pollutants into the building through any cracks and crevices available.
Substructures/slab penetrations	Radon and other soil gases and moisture laden air or microbial contaminated air often travel through crawl spaces and other substructures into the building.

gases and materials that are automatically top priorities for the industrial hygienist to investigate and mitigate) that are considered extremely hazardous. These too garner the industrial hygienist's immediate attention and remedial action. In this text, we focus on asbestos, silica, formaldehyde, and lead as those hazardous contaminants (keeping in mind that there are others) requiring the immediate attention of the industrial hygienist.

ASBESTOS EXPOSURE (OSHA, 2003)

Asbestos is the name given to a group of naturally occurring minerals widely used in certain products, such as building materials and vehicle brakes, to resist heat and corrosion. Asbestos includes chrysotile, amosite, crocidolite, tremolite asbestos, anthophyllite asbestos, actinolite asbestos, and any of these materials that have been chemically treated or altered. Typically, asbestos appears as a whitish, fibrous material that may release fibers that range in texture from coarse to silky; however, airborne fibers that can cause health damage may be too small to be seen with the naked eye.

An estimated 1.3 million employees in construction and general industry face significant asbestos exposure on the job. Heaviest exposures occur in the construction industry, particularly during the removal of asbestos during renovation or demolition (abatement). Employees are also likely to be exposed during the manufacture of asbestos products (such as textiles, friction products, insulation, and other building materials) and automotive brake and clutch repair work.

The inhalation of asbestos fibers by workers can cause serious diseases of the lungs and other organs that may not appear until years after the exposure has occurred. For instance, asbestosis can cause a buildup of scar-like tissue in the lungs and result in loss of lung function that often progresses to disability and death. As mentioned, asbestos fibers associated with these health risks are too small to be seen with the naked eye, and smokers are at higher risk of developing some asbestos-related diseases. For example, exposure to asbestos can cause asbestosis (scarring of the lungs resulting in loss of lung function that often progresses to disability and death); mesothelioma (cancer affecting the membranes lining the lungs and abdomen); lung cancer; and cancers of the esophagus, stomach, colon, and rectum.

The Occupational Safety and Health Administration (OSHA) has issued the following three standards to assist industrial hygienists with compliance and to protect workers from exposure to asbestos in the workplace:

- 29 CFR 1926.1101 covers construction work, including alteration, repair, renovation, and demolition of structures containing asbestos.
- 29 CFR 1915.1001 covers asbestos exposure during work in shipyards.
- 29 CFR 1910.1001 applies to asbestos exposure in general industry, such as exposure during brake and clutch repair, custodial work, and manufacture of asbestos-containing products.

The standards for the construction and shipyard industries classify the hazards of asbestos work activities and prescribe particular requirements for each classification:

- Class I—Is the most potentially hazardous class of asbestos jobs and involves the removal of thermal system insulation and sprayed-on or troweled-on surfacing asbestos-containing materials or presumed asbestos-containing materials.
- Class II—Includes the removal of other types of asbestos-containing materials that are not thermal system insulation, such as resilient flooring and roofing materials containing asbestos.
- Class III—Focuses on repair and maintenance operations where asbestos-containing materials are disturbed.
- Class IV—Pertains to custodial activities where employees clean up asbestos-containing waste and debris.

There are equivalent regulations in states with OSHA-approved state plans.

Permissible Exposure Limits

Employee exposure to asbestos must not exceed 0.1 fiber per cubic centimeter (f/cc) of air, averaged over an 8-hour work shift. Short-term exposure must also be limited to not more than 1 f/cc, averaged over 30 minutes. Rotation of employees to achieve compliance with either permissible exposure limit (PEL) is prohibited.

Exposure Monitoring

In construction and shipyard work, unless the industrial hygienist is able to demonstrate that employee exposures will be below the PELs (a negative exposure assessment), it is generally a requirement that monitoring for workers in class I and II regulated areas be conducted. For workers in other operations where exposures are expected to exceed one of the PELs, periodic monitoring must be conducted. In general industry, for workers who may be exposed above a PEL or above the excursion limit, initial monitoring must be conducted. Subsequent monitoring at reasonable intervals must be conducted, and in no case at intervals greater than 6 months for employees exposed above a PEL.

Competent Person

In all operations involving asbestos removal (abatement), employers must name a competent person qualified and authorized to ensure worker safety and health, as required by Subpart C, "General Safety and Health Provisions for Construction" (29 CFR 1926.20). Under the requirements for safety and health prevention programs, the competent person must frequently inspect job sites, materials, and equipment. A fully trained and licensed industrial hygienist often fills this role.

In addition, for class I jobs the competent person must inspect on-site at least once during each work shift and upon employee request. For class II and III jobs, the competent person must inspect often enough to assess changing conditions and upon employee request.

Regulated Areas

In general industry and construction, regulated areas must be established where the 8-hour time-weighted average (TWA) or 30-minute excursions values for airborne asbestos exceed the PELs. Only authorized persons wearing appropriate respirators can enter a regulated area. In regulated areas, eating, smoking, drinking, chewing tobacco or gum, and applying cosmetics are prohibited. Warning signs must be displayed at each regulated area and must be posted at all approaches to regulated areas.

Methods of Compliance

In both general industry and construction, employers must control exposures using engineering controls, to the extent feasible. Where engineering controls are not feasible to meet the exposure limit, they must be used to reduce employee exposures to the lowest levels attainable and must be supplemented by the use of respiratory protection.

Respirators

In general industry and construction, the level of exposure determines what type of respirator is required; the standards specify the respirator to be used. Keep in mind that respirators must be used during all class I asbestos jobs. Refer to 29 CFR 1926.103 for further guidance on when respirators must be worn.

Labels

Caution labels must be placed on all raw materials, mixtures, scrap, waste, debris, and other products containing asbestos fibers.

Protective Clothing

For any employee exposed to airborne concentrations of asbestos that exceed the PEL, the employer must provide and require the use of protective clothing such as coveralls or similar full-body clothing, head coverings, gloves, and foot coverings. Wherever the possibility of eye irritation exists, face shields, vented goggles, or other appropriate protective equipment must be provided and worn.

Training

For employees involved in each identified work classification, training must be provided. The specific training requirements depend upon the particular class of work being performed. In general industry, training must be provided to all employees exposed above a PEL. Asbestos awareness training must also be provided to employees who perform housekeeping operations covered by the standard. Warning labels must be placed on all asbestos products, containers, and installed construction materials when feasible.

Recordkeeping

The employer must keep an accurate record of all measurements taken to monitor employee exposure to asbestos. This record is to include the date of measurement; operation involving exposure; sampling and analytical methods used, and evidence of their accuracy; number, duration, and results of samples taken; type of respiratory protective devices worn; and name, social security number, and the results of all employee exposure measurements. This record must be kept for 30 years.

Hygiene Facilities and Practices

Clean change rooms must be furnished by employers for employees who work in areas where exposure is above the TWA or excursion limit. Two lockers or storage facilities must be furnished and separated to prevent contamination of the employee's street clothes from protective work clothing and equipment. Showers must be furnished so that employees may shower at the end of the work shift. Employees must enter and exit the regulated area through the decontamination area.

The equipment room must be supplied with impermeable, labeled bags and containers for the containment and disposal of contaminated protective clothing and equipment.

Lunchroom facilities for those employees must have a positive-pressure, filtered air supply and be readily accessible to employees. Employees must wash their hands and face prior to eating, drinking, or smoking. The employer must ensure that employees do not enter lunchroom facilities with protective work clothing or equipment unless surface fibers have been removed from the clothing or equipment.

Employees may not smoke in work areas where they are occupationally exposed to asbestos.

Medical Exams

In general industry, exposed employees must have a preplacement physical examination before being assigned to an occupation exposed to airborne concentrations of asbestos at or above the action level or excursion level. The physical examination must include chest x-ray, medical and work history, and pulmonary function tests. Subsequent exams must be given annually and upon termination of employment, though chest x-rays are required annually only for older workers whose first asbestos exposure occurred more than 10 years ago.

In construction, examinations must be made available annually for workers exposed above the action level or excursion limit for 30 or more days per year or who are required to wear negative-pressure respirators; chest x-rays are at the discretion of the physician.

SILICA EXPOSURE

Crystalline silica (SiO_2) is a major component of the earth's crust. In pure, natural form, SiO_2 crystals are minute, very hard, translucent, and colorless. Most mined minerals contain some SiO_2. *Crystalline* refers to the orientation of SiO_2 molecules in a fixed pattern as opposed to a nonperiodic, random molecular arrangement defined as amorphous (e.g., diatomaceous earth). Therefore, silica exposure occurs in a wide variety of settings, such as mining, quarrying, and stone-cutting operations; ceramics and vitreous enameling; and in use of filters for paints and rubber. The wide use and multiple applications of silica in industrial applications combine to make silica a major occupational health hazard (silicosis), which can lead to death.

Silicosis is a disabling, nonreversible, and sometimes fatal lung disease caused by overexposure to respirable crystalline silica. More than 1 million U.S. workers are exposed to crystalline silica, and each year more than 250 die from silicosis (see Table 17.7). There is no cure for the disease, but it is 100% preventable if employers, workers, and health professionals work together to reduce exposures.

Guidelines for Control of Occupational Exposure to Silica

In accordance with OSHA's standard for air contaminants (29 CFR 1910.1000), employee exposure to airborne crystalline silica shall not exceed an 8-hour time-weighted average limit (variable) as stated in 29 CFR 1910.1000, Table Z-3, or a limit set by a state agency whenever a state-administered Occupational Safety and Health Plan is in effect.

As mandated by OSHA, the first mandatory requirement is that employee exposure be eliminated through the implementation of feasible engineering controls (e.g., dust suppression and ventilation). After all such controls are implemented and they

TABLE 17.7
Deaths from Silica in the Workplace

Occupation	PMR
Miscellaneous metal and plastic machine operators	168.44
Hand molders and shapers, except jewelers	64.12
Crushing and grinding machine operators	50.97
Hand molding, casting, and forming occupations	35.70
Molding and casting machine operators	30.60
Mining machine operators	19.61
Mining occupations (not elsewhere classified)	15.33
Construction trades (not elsewhere classified)	14.77
Grinding, abrading, buffing, and polishing machine operators	8.47
Heavy equipment mechanics	7.72
Miscellaneous material moving equipment operators	6.92
Millwrights	6.56
Crane and tower operators	6.02
Brickmasons and stonemasons	4.71
Painters, construction, and maintenance	4.50
Furnace, kiln, oven operators, except food	4.10
Laborers, except construction	3.79
Operating engineers	3.56
Welders and cutters	3.01
Machine operators, not specified	2.86
Not specified mechanics and repairers	2.84
Supervisors, production occupations	2.73
Construction laborers	2.14
Machinists	1.79
Janitors and cleaners	1.78

Industry	
Metal mining	69.51
Miscellaneous nonmetallic mineral and stone products	55.31
Nonmetallic mining and quarrying, except fuel	49.77
Iron and steel foundries	31.15
Pottery and related products	30.73
Structural clay products	27.82
Coal mining	9.26
Blast furnaces, steelworks, rolling and finishing mills	6.49
Miscellaneous fabricated metal products	5.87
Miscellaneous retail stores	4.63
Machinery, except electrical (not elsewhere classified)	3.96
Other primary metal industries	3.63

(Cont.)

TABLE 17.7
(Cont.)

Occupation	PMR
Industrial and miscellaneous chemicals	2.72
Not specified manufacturing industries	2.67
Construction	1.82

Source: National Institute for Occupational Safety and Health, U.S. Department of Health and Human Services, Public Health Service, Centers for Disease Control and Prevention, *Work-Related Lung Disease Surveillance Report*, DHHA (NIOSH) Publication 96–134, Table 3–8, 2002.

Note: The PMR column is the observed number of deaths from silicosis per occupation divided by the expected number of deaths. Therefore, a value of 1 indicates no additional risk. A value of 10 would indicate a risk ten times greater than the normal risk of silicosis.

do not control to the permissible exposure, each employer must rotate its employees to the extent possible in order to reduce exposure. Only when all engineering or administrative controls have been implemented, and the level of respirable silica still exceeds permissible exposure limits, may an employer rely on a respirator program pursuant to the mandatory requirements of 29 CFR1910.134. Generally, where working conditions or other practices constitute recognized hazards likely to cause death or serious physical harm, they must be corrected.

FORMALDEHYDE (HCHO) EXPOSURE

Formaldehyde (HCHO) is a colorless, flammable gas with a pungent suffocating odor. Formaldehyde is common to the chemical industry. It is the most important aldehyde produced commercially, and is used in the preparation of urea-formaldehyde and phenol-formaldehyde resins. It is also produced during the combustion of organic materials and is a component of smoke.

The major sources in workplace settings are in manufacturing processes (used in the paper, photographic, and clothing industries) and building materials. Building materials may contain phenol, urea, thiourea, or melamine resins that contain HCHO. Degradation of HCHO resins can occur when these materials become damp from exposure to high relative humidity, or if the HCHO materials are saturated with water during flooding, or when leaks occur. The release of HCHO occurs when the acid catalysts involved in the resin formulation are reactivated. When temperatures and relative humidity increase, outgassing increases (DOH, 2003).

Formaldehyde exposure is most common through gas-phase inhalation. However, it can also occur through liquid-phase skin absorption. Workers can be exposed during direct production, treatment of materials, and production of resins. Health care professionals, pathology and histology technicians, and teachers and students who handle preserved specimens are potentially at high risk.

Studies indicate that formaldehyde is a potential human carcinogen. Airborne concentrations above 0.1 ppm can cause irritation of the eyes, nose, and throat. The severity of irritation increases as concentrations increase; at 100 ppm it is immediately dangerous to life and health. Dermal contact causes various skin reactions, including sensitization, which might force persons thus sensitized to find other work.

OSHA requires that the employer conduct initial monitoring to identify all employees who are exposed to formaldehyde at or above the action level or Short Term Exposure Limit (STEL) and to accurately determine the exposure of each employee so identified. If the exposure level is maintained below the STEL and the action level, employers may discontinue exposure monitoring, until there is such a change that could affect exposure levels. The employer must also monitor employee exposure promptly, upon receiving reports of formaldehyde-related signs and symptoms.

In regards to exposure control, the best prevention is provided by source control (if possible). The selection of HCHO-free or low-emitting products such as exterior-grade plywood, which uses phenol HCHO resins, for indoor use is the best starting point.

Secondary controls include filtration, sealants, and fumigation treatments. Filtration can be achieved using selected adsorbents. Sealants involve coating the materials in question with two or three coats of nitrocellulose varnish or water-based polyurethane. Three coats of these materials can reduce outgassing by as much as 90%.

Training is required at least annually for all employees exposed to formaldehyde concentrations of 0.1 ppm or greater. The training will increase employees' awareness of specific hazards in their workplace and of the control measures employed. The training also will assist successful medical surveillance and medical removal programs. These provisions will only be effective if employees know what signs or symptoms are related to the health effects of formaldehyde, if they are periodically encouraged to do so.

LEAD EXPOSURE

Lead has been poisoning workers for thousands of years. Most occupational overexposures to lead have been found in the construction trades, such as plumbing, welding, and painting. In plumbing, soft solder (banned for many uses in the United States), used chiefly for soldering tinplate and copper pipe joints, is an alloy of lead and tin. Although the use of lead-based paint in residential applications has been banned, since lead-based paint inhibits the rusting and corrosion of iron and steel, it is still used in construction projects. Significant lead exposures can also arise from removing paint from surfaces previously coated with lead-based paint. According to OSHA 93-47 (2003), the operations that generate lead dust and fume include the following:

- Flame-torch cutting, welding, the use of heat guns, sanding, scraping and grinding of lead-painted surfaces in repair, reconstruction, dismantling, and demolition work
- Abrasive blasting of structures containing lead-based paints
- Use of torches and heat guns, and sanding, scraping, and grinding of lead-based paint surfaces during remodeling or abating lead-based paint
- Maintaining process equipment or exhaust ductwork

Health Effects of Lead

There are several routes of entry in which lead enters the body. When absorbed into the body in certain doses, lead is a toxic substance. Lead can be absorbed into the body by inhalation and ingestion. Except for certain organic lead compounds not covered by OSHA's lead standard (29 CFR 1926.62), such as tetraethyl lead, lead, when scattered in the air as a dust, fume, or mist, can be absorbed into the body by inhalation.

A significant portion of the lead that can be inhaled or ingested gets into the bloodstream. Once in the bloodstream, lead is circulated throughout the body and stored in various organs and tissues. Some of this lead is quickly filtered out of the body and excreted, but some remains in the blood and other tissues. As exposure to lead continues, the amount stored in the body will increase if more lead is being absorbed than is being excreted. Cumulative exposure to lead, which is typical in construction settings, may result in damage to the blood, nervous system, kidneys, bones, heart, and reproductive system and contributes to high blood pressure. Some of the symptoms of lead poisoning include the following:

- Poor appetite
- Dizziness
- Pallor
- Headache
- Irritability/anxiety
- Constipation
- Sleeplessness
- Weakness
- Insomnia
- "Lead line" in gums
- Fine tremors
- Hyperactivity
- Wrist drop (weakness of extensor muscles)
- Excessive tiredness
- Numbness
- Muscle and joint pain or soreness
- Nausea
- Reproductive difficulties

Lead Standard Definitions

According to OSHA's lead standard, the terms listed below have the following meanings:

- *Action level*—Employee exposure, without regard to the use of respirators, to an airborne concentration of lead of 30 micrograms per cubic meter of air (30 $\mu g/m^3$), averaged over an 8-hour period.

- *Permissible exposure limit* (PEL)—Concentration of airborne lead to which an average person may be exposed without harmful effects. OSHA has established a PEL of 50 micrograms per cubic meter of air (50 µg/m³), averaged over an 8-hour period. If an employee is exposed to lead for more than 8 hours in any workday, the permissible exposure limit, a time-weighted average (TWA) for that day, shall be reduced according to the following formula:

Maximum permissible limit (in µg/m³) = 400 × hours worked in the day

- When respirators are used to supplement engineering and administrative controls to comply with the PEL and the requirements of the lead standard's respiratory protection rules, employee exposure, for the purpose of determining whether the employer has complied with the PEL, may be considered to be at the level provided by the protection factor of the respirator for those periods the respirator is worn. Those periods may be averaged with exposure levels during periods when respirators are not worn to determine the employee's daily TWA exposure.
- *µg/m³*—Micrograms per cubic meter of air. A microgram is 1 millionth of a gram. There are 454 grams in a pound.

Worker Lead Protection Program

The employer is responsible for the development and implementation of a worker lead protection program. This program is essential in minimizing worker risk of lead exposure. The most effective way to protect workers is to minimize exposure through the use of engineering controls and good work practices.

At the minimum, the following elements should be included in the employer's worker protection program for employees exposed to lead:

- Hazard determination, including exposure assessment
- Engineering and work practice controls
- Respiratory protection
- PPE (protective work clothing and equipment)
- Housekeeping
- Hygiene facilities and practices
- Medical surveillance and provisions for medical removal
- Employee information and training
- Signs
- Recordkeeping

MOLD CONTROL

Molds can be found almost anywhere; they can grow on virtually any organic substance, as long as moisture and oxygen are present. The earliest known writings that appear to discuss mold infestation and remediation (removal, cleaning up) are found in Leviticus 14, Old Testament.

WHERE ARE MOLDS TYPICALLY FOUND?

Molds have been found growing in office buildings, schools, automobiles, private homes, and other locations where water and organic matter are left unattended. Mold is not a new issue—just one that, until recently, has received little attention by regulators in the United States. That is, there are no state or federal statutes or regulations regarding molds and IAQ.

Molds reproduce by making spores that usually cannot be seen without magnification. Mold spores waft through the indoor and outdoor air continually. When mold spores land on a damp spot indoors, they may begin growing and digesting whatever they are growing on in order to survive. Molds generally destroy the things they grow on (USEPA, 2001).

The key to limiting mold exposure is to prevent the germination and growth of mold. Since mold requires water to grow, it is important to prevent moisture problems in buildings. Moisture problems can have many causes, including uncontrolled humidity. Some moisture problems in workplace buildings have been linked to changes in building construction practices during the 1970s, 1980s, and 1990s. Some of these changes have resulted in buildings that are tightly sealed, but may lack adequate ventilation, potentially leading to moisture buildup. Building materials, such as drywall, may not allow moisture to escape easily. Moisture problems may include roof leaks, landscaping or gutters that direct water into or under the building, and unvented combustion appliances. Delayed maintenance or insufficient maintenance are also associated with moisture problems in buildings. Moisture problems in temporary structures have frequently been associated with mold problems.

Building maintenance personnel, architects, and builders need to know effective means of avoiding mold growth that might arise from maintenance and construction practices. Locating and cleaning existing growths are also paramount to decreasing the health effects of mold contamination. Using proper cleaning techniques is important because molds are incredibly resilient and adaptable (Davis, 2001).

Molds can elicit a variety of health responses in humans. The extent to which an individual may be affected depends upon his or her state of health, susceptibility to disease, the organisms with which he or she came in contact, and the duration and severity of exposure (Ammann, 2000). Some people experience temporary effects that disappear when they vacate infested areas (Burge, 1997). In others, the effects of exposure may be long term or permanent (Yang, 2001).

It should be noted that systemic infections caused by molds are not common. Normal, healthy individuals can resist systemic infection from airborne molds.

Those at risk for system fungal infection are severely immunocompromised individuals such as those with HIV/AIDs, individuals who have had organ or bone marrow transplants, and persons undergoing chemotherapy.

In 1994, an outbreak of *Stachybotrys chartarum* in Cleveland, Ohio, was believed by some to have caused pulmonary hemorrhage in infants. Sixteen of the infants died. The CDC sponsored a review of the cases and concluded that the scientific evidence provided did not warrant the conclusion that inhaled mold was the cause of the illnesses in the infants. However, the panel also stated that further research was warranted, as the study design for the original research appeared to be flawed (CDC, 1999).

Below is a list of mold components known to elicit a response in humans.

- *Volatile organic compounds* (VOCs)—"Molds produce a large number of volatile organic compounds. These chemicals are responsible for the musty odors produced by growing molds" (McNeel and Kreutzer, 1996, p. 66). VOCs also provide the odor in cheeses, and the off taste of mold-infested foods. Exposure to high levels of volatile organic compounds affects the central nervous system (CNS), producing such symptoms as headaches, attention deficit, inability to concentrate, and dizziness (Ammann, 2000). According to McNeel and Kreutzer, at present the specific contribution of mold volatile organic compounds in building-related health problems has not been studied. Also, mold volatile organic compounds are likely responsible for only a small fraction of total VOCs indoors (Davis, 2001).
- *Allergens*—All molds, because of the presence of allergens on spores, have the potential to cause an allergic reaction in susceptible humans (Rose, 1999). Allergic reactions are believed to be the most common exposure reaction to molds. These reactions can range from mild, transitory response, like runny eyes, runny nose, throat irritation, coughing, and sneezing, to severe, chronic illnesses such as sinusitis and asthma (Ammann, 2000).
- *Mycotoxins*—Natural organic compounds that are capable of initiating a toxic response in vertebrates (McNeel and Kreutzer, 1996). Some molds are capable of producing mycotoxins. Molds known to potentially produce mycotoxins and which have been isolated in infestations causing adverse health effects include certain species of *Acremonium, Alternaria, Aspergillus, Chaetomium, Caldosporium, Fusarium, Paecilomyces, Penicillium, Stachybotrys*, and *Trichoderma* (Yang, 2001).

While a certain type of mold or mold strain may have the genetic potential for producing mycotoxins, specific environmental conditions are believed to be needed for the mycotoxins to be produced. In other words, although a given mold might have the potential to produce mycotoxins, it will not produce them if the appropriate environmental conditions are not present (USEPA, 2001).

Currently, the specific conditions that cause mycotoxin production are not fully understood. The USEPA recognizes that mycotoxins have a tendency to concentrate in fungal spores, and that there is limited information currently available regarding the process involved in fungal spore release. As a result, the USEPA is currently conducting research in an effort to determine "the environmental conditions required for sporulation, emission, aerosolization, dissemination and transport of [*Stachybotrys*] into the air" (USEPA, 2001).

MOLD PREVENTION

As mentioned, the key to mold control is moisture control. Solve moisture problems before they become mold problems. Several mold prevention tips are listed below.

- Fix leaky plumbing and leaks in the building envelope as soon as possible.
- Watch for condensation and wet spots. Fix sources of moisture problems as soon as possible.
- Prevent moisture due to condensation by increasing surface temperature or reducing the moisture level in air (humidity). To increase surface temperature, insulate or increase air circulation. To reduce the moisture level in air, repair leaks, increase ventilation (if outside air is cold and dry), or dehumidify (if outdoor air is warm and humid).
- Keep heating, ventilation, and air conditioning (HVAC) drip pans clean, flowing properly, and unobstructed.
- Vent moisture-generating appliances, such as dryers, to the outside where possible.
- Perform regular building/HVAC inspections and maintenance as scheduled.
- Maintain low indoor humidity, below 60% relative humidity (RH), ideally 30–50%, if possible.
- Clean and dry wet or damp spots within 48 hours.
- Do not let foundations stay wet. Provide drainage and slope the ground away from the foundation.

MOLD REMEDIATION

At the present time, there are no standardized recommendations for mold remediation; however, USEPA is working on guidelines. There are certain aspects of mold cleanup, however, that are agreed upon by many practitioners in the field:

- A commonsense approach should be taken when assessing mold growth. For example, it is generally believed that small amounts of growth, like those commonly found on shower walls, pose no immediate health risk to most individuals.
- Persons with respiratory problems, a compromised immune system, or fragile health should not participate in cleanup operations.
- Cleanup crews should be properly attired. Mold should not be allowed to touch bare skin. Eyes and lungs should be protected from aerosol exposure.
- Adequate ventilation should be provided while at the same time containing the infestation in an effort to avoid spreading mold to other areas.
- The source of moisture must be stopped and all areas infested with mold thoroughly cleaned. If thorough cleaning is not possible due to the nature of the material (porous vs. semi- and nonporous), all contaminated areas should be removed.

Safety tips that should be followed when remediating moisture and mold problems include:

- Do not touch mold or moldy items with bare hands.
- Do not get mold or mold spores in your eyes.
- Do not breathe in mold or mold spores.
- Consider using PPE when disturbing mold. The minimum PPE is a respirator, gloves, and eye protection.

MOLD CLEANUP METHODS

A variety of mold cleanup methods are available for remediating damage to building materials and furnishings caused by moisture control problems and mold growth. These include wet vacuum, damp wipe, HEPA vacuum, and the removal and sealing of damaged materials in plastic bags. The specific method or group of methods used will depend on the type of material affected.

A COTTAGE INDUSTRY IS BORN

As a result of the Legionnaires' incident in Philadelphia, the WHO study, and significant amounts of media attention, Legionnaires' disease, sick building syndrome (SBS) and indoor air pollution became common terms used by common working people in the workplace. An offshoot of this media attention was a new cottage industry consisting of so-called experts in indoor air pollution who commenced selling their services to conduct sick building surveys. From about 1985 until the early 1990s this new industry was booming. Since then, many of these new enterprises folded their operations because of a lack of business; the initial scare wore off.

During its heyday, the movement to solve the sick building syndrome problem resulted in some air pollutants being identified as the culprits or potential culprits in causing SBS. One such category of air pollutants in nonindustrial environments was identified as volatile organic compounds (VOCs). In 1989 the WHO classified (according to boiling-point ranges) the entire range of organic pollutants into four groups: (1) very volatile (gaseous compounds), (2) volatile, (3) semivolatile, and (4) organic compounds associated with particulate matter.

REFERENCES AND RECOMMENDED READING

American Conference of Governmental Industrial Hygienists (ACGIH). 1997. *Guidelines for assessment and control of aerosols.* Cincinnati: ACGIH.

American Industrial Hygiene Association (AIHA). 1989. *Practitioners approach to indoor air quality investigations.* Fairfax, VA: AIHA.

American Industrial Hygiene Association (AIHA). 1996. *Field guide for the determination of biological contaminants in environmental samples.* Fairfax, VA: AIHA.

Ammann, H. 2000. Is indoor mold contamination a threat? Washington State Department of Health. http://www.doh.wa.gov/ehp/ocha/mold.html (accessed August 9, 2003).

Burge, H. A. 1997. The fungi: How they grow and their effects on human health. *Heating/Piping/Air Conditioning*, July, pp. 69–75.

Burge, H. A., and Hoyer, M. E. 1998. Indoor air quality. In DiNardi, S. R., ed., *The occupational environment—Its evaluation and control.* Fairfax, VA: American Industrial Hygiene Association.

Byrd, R. R. 2003. *IAQ FAG part 1.* Glendale, CA: Machado Environmental Corporation.

CDC. 1999. *Reports of members of the CDC External Expert Panel on Acute Idiopathic Pulmonary Hemorrhage in Infants: A synthesis.* Washington, DC: Centers for Disease Control.

Davis, P. J. 2001. *Molds, toxic molds, and indoor air quality.* Sacramento, CA: California Research Bureau, California State Library.

Di Tommaso, R. M., Nino, E., and Fracastro, G. V. 1999. Influence of the boundary thermal conditions on the air change efficiency indexes. *Indoor Air* 9:63–69.

Dockery, D. W., and Spengler, J. D. 1981. Indoor-outdoor relationships of respirable sulfates and particles. *Atmos. Environ.* 15:335–43.

DOH. 2003. Formaldehyde. Washington State Department of Health, Office of Environmental Health and Safety. http://www.doh.wa.gov/ehp/ts/IAQ/Formaldehyde.HTM (accessed August 2003).

Hollowell, C. D., et al. 1979a. Impact of infiltration and ventilation on indoor air quality. *ASHRAE J.*, July, pp. 49–53.

Hollowell, C. D., et al. 1979b. Impact of energy conservation in buildings on health. In Razzolare, R. A., and Smith, C. B., eds., *Changing energy use futures.* New York: Pergamon.

Leviticus 14.

McNeel, S., and Kreutzer, R. 1996. Fungi and indoor air quality. *Health and Environment Digest* 10(2).

Molhave, L. 1986. Indoor air quality in relation to sensory irritation due to volatile organic compounds. *ASHRAE Trans.* 92:306–16.

OSHA. 2003. Lead exposure in construction. OSHA 93-47, U.S. Department of Labor, Occupational Safety and Health Administration. http://www.osha-slc.gov/pls/oshaweb/owadisp.show_document?p_table=FACT_Sheet (accessed August 2007).

Passon, L., et al. 1996. Sick-building syndrome and building-related illnesses. *Med. Lab. Observer* 28:84–95.

Peng, S. H., and Davidson, I. 1999. Performance evaluation of a displacement ventilation system for improving indoor air quality: A numerical study. Paper presented at the 8th International Conference on Indoor Air Quality and Climate, Edinburgh, Scotland.

Rose, C. F. 1999. Antigens. In *ACGIH bioaerosols assessment and control*, chap. 25, pp. 25-1 to 25-11. Cincinnati, OH: ACGIH.

Spellman, F. R., and Whiting, N. 2006. *Environmental science and technology: Concepts and applications.* 2nd ed. Rockville, MD: Government Institutes.

Spengler, J. D., et al. 1979. Sulfur dioxide and nitrogen dioxide levels inside and outside homes and the implications on health effects research. *Environ. Sci. Technol.* 13:1276–80;

USEPA. 1986. A citizen's guide to radon.

USEPA. 1987. Indoor air facts no. 1, EPA and indoor air quality.

USEPA. 1988. *The inside story—A guide to indoor air quality.* Washington, DC: Author.

USEPA. 1990. Residential air cleaners indoor air facts no. 7. Air and radiation.

USEPA. 1991. *Building air quality: A guide for building owners and facility managers.* Washington, DC: Author.

USEPA. 2001. Indoor air facts no. 4 (revised) sick building syndrome. http://www.epa.gov/iaq/pubs/sbs.html (accessed August 9, 2007).

USEPA. 2005. Indoor environmental management branch. Children's health initiative: Toxic mold. http://epa.gov.appedwww/crb/iemb/child.htm (accessed January 9, 2008).

Wadden, R. A., and Scheff, P. A. 1983. *Indoor air pollution: Characteristics, predications, and control.* New York: John Wiley & Sons.

Woods, J. E. 1980. Environmental implications of conservation and solar space heating. BEUL 80-3, Engineering Research Institute, Iowa State University, Ames.

World Health Organization (WHO). 1982. *Indoor air pollutants, exposure and health effects assessment. Euro-reports and Studies no. 78: Working group report.* Copenhagen: WHO Regional Office.

World Health Organization (WHO). 1984. *Indoor air quality research Euro-reports and studies no. 103.* Copenhagen: WHO Regional Office for Europe.

Yang, C. S. 2001. Toxic effects of some common indoor fungi. http://www.envirovillage. com/Newsletters/Enviros/No4_09.htm (accessed August 6, 2006).

Zhang, Y., Wang, X., Riskowski, G. L., and Christianson, L. L. 2001. Quantifying ventilation effectiveness for air quality control. *Transactions of the American Society of Agricultural and Biological engineers* (ASABE) 44:385–90.

Index

A

Absolute humidity, 126
Absolute pressure, 10, 61
Absolute temperature, 36
Absolute viscosity, 72
Absorption-based pollution control, 230–233
Acid, defined, 10
Acidic solution, 10
Acid precipitation, 10, 210, 211
   regulation, 186–187
Acid rain, 186, *See also* Acid precipitation
Acids and bases, 10, 104–105
Acid surge, 10
Activated carbon, 233
Acute effects, 253
Adiabatic expansion and cooling, 201–202
Adiabatic lapse rate, 71–72
Adsorption-based pollution control, 230,
   233–234
Advection, 174
Advective winds, 135
Aerobic respiration, 5–6
Aerodynamic equivalent diameter, 81–82
Aerosols, 28
Afterburners, 237–238
Air, 3, 15
   ambient, 198
   defining, 4
   life and, 3
   local mean age of, 262
   origin of, 4–5
   oxygen and, 15
   purity of, 4
   scientific discoveries, 16–17
Airborne toxins, 10
Air change efficiency (ACE), 262
Air composition, 4–5, 16, 115
   component properties, 17–27, *See also specific*
     components
   unpolluted air, 197
Air conditioning systems, 29
Air currents, 133, 146
Air diffusion performance index (ADPI), 263
Air masses, 146–147
Air movement, 133–142, *See also* Winds
   air currents, 133, 146
   causes, 134–135

convection, 146
Coriolis effect, 136–137
drag effects, 137–138
fronts, 146–147
gas laws and, 135–136
human comfort and, 138–139
jet streams, 140, 145
plume behavior, 204–206
pollutant dispersion and, *See* Dispersion
thermal circulation, 136
topography and, 203
turbulence, 200
Air parcels, 200–201
Air pollutant, defined, 197, *See also* Air
     pollution; Pollutants; *specific pollutants*
Air pollution, 197–218
   air movement and, 146
   air quality, 173, 178–179, *See also* Air quality
     management
   Bhopal disaster, 193–194
   dispersion and deposition, *See* Dispersion
   health effects, 179, 197
   indoor air quality problems, 249–252, *See also*
     Indoor air quality
   mortality, 8
   natural sources, 197, 198
   NIMBY issues, 222
   photochemical smog, 11, 110
   pollutants, 197, 209–218, *See also specific*
     pollutants
   precursor pollutants, 209
   regulation, *See* Air quality management
   sources, 8–9
   thermal inversions and, 147–149, 203–204
   vapor plume-induced icing, 193
   visibility impairment effects, 192
Air Pollution Control Act of 1955, 182
Air pollution control technologies, 219–242
   choices, 220–223
   comparison of, 231
   dry particulates, 224–230, *See also* Particulate
     control technologies
   factors in selecting, 224
   gaseous pollutants, 230–243, *See also* Gaseous
     pollutant control technologies
   gas treatment (residence) time, 70
   gas velocities in, 69–70
   mobile sources, 239–242